Modern Birkhäuser Classics

Many of the original research and survey monographs in pure and applied mathematics published by Birkhäuser in recent decades have been groundbreaking and have come to be regarded as foundational to the subject. Through the MBC Series, a select number of these modern classics, entirely uncorrected, are being re-released in paperback (and as eBooks) to ensure that these treasures remain accessible to new generations of students, scholars, and researchers.

T0184494

The Versatile Soliton

Alexandre T. Filippov

Reprint of the 2000 Edition

Birkhäuser
Boston • Basel • Berlin

Alexandre T. Filippov
Bogoliubov Laboratory of Theoretical Physics
Joint Institute for Nuclear Research
Joliot Curie 6
RU-141980 Dubna
Russia

ISBN 978-0-8176-4973-9   e-ISBN 978-0-8176-4974-6
DOI 10.1007/978-0-8176-4974-6
Springer New York Dordrecht Heidelberg London

Library of Congress Control Number: 2010927097

Mathematics Subject Classification (2010): 00Axx, 00A79, 35Q51, 76D33

Printed on acid-free paper

Birkhäuser is part of Springer Science+Business Media (www.birkhauser.com)

Alexandre T. Filippov

The Versatile Soliton

Birkhäuser
Boston • Basel • Berlin

Alexandre T. Filippov
Bogoliubov Laboratory of Theoretical Physics
Joint Institute for Nuclear Research
Joliot Curie 6
RU-141980 Dubna
Russia

Library of Congress Cataloging-in-Publication Data

Filippov, A. T. (Alexandr Tikhonovich)
   [Mnogolikii soliton. English]
   The versatile soliton / Alexandre T. Filippov.
     p.   cm.
   Includes bibliographical references and index.
   ISBN 0-8176-3635-8 (alk. paper). — ISBN 3-7643-3635-8 (alk. paper)
   1. Solitons. I. Title.
  QC174.26.W28 .F5513   2000
  530.12'4–dc21
                                       00-036770
                                          CIP

AMS Subject Classifications: 00Axx, 00A79, 35Q51, 76D33

Printed on acid-free paper.
© 2000 Birkhäuser Boston

Birkhäuser

ISBN 0-8176-3635-8   SPIN 10464272
ISBN 3-7643-3635-8

Reformatted from author's files in LATEX 2e by TEXniques, Inc., Cambridge, MA.
Printed and bound by Hamilton Printing Company, Rensselaer, NY.
Printed in the United States of America.

9 8 7 6 5 4 3 2 1

Contents

II   Nonlinear Oscillations and Waves   59

III   The Present and Future of the Soliton   129

Introduction

The temple of science is a multi-faceted building

Albert Einstein (1879–1955)

We all have an intuitive familiarity with the closely related notions of *oscillation*, *vibration* and *wave*. Any periodic motion of any object, or periodic change of its state, is called an *oscillation* or *vibration*; examples are evident in our everyday life: our pulse, the swinging of a pendulum, ticking of a clock, and so on. The scientific approach to oscillations starts by neglecting physical differences between vibrating objects, by extracting features that are common to all oscillations, and then finding mathematical laws to describe these oscillations.

Oscillations and vibrations usually generate waves. Waves on water are easily recognized as physical phenomena, but radio waves can only be imagined, making it more difficult to establish their physical reality. To imagine radio waves, one has to know some physics and mathematics. Since the theory of waves is concerned with characteristics that are common to all waves, independent of their physical nature, it is sufficient to comprehend just one example, for example, waves on water. However, we should note that the wave on water is a much more complex phenomenon than is apparent to the eye. So, in this book we will first examine much simpler waves.

In general, a *wave* may be defined as a progression through matter of a state of motion. Waves have a close relationship with oscillations. In fact, the simplest *periodic wave* in any medium is a progression of oscillatory motion to neighboring points of the medium. Such waves are usually emitted by vibrating objects, e.g., sound waves generated by a tuning fork. The vibrating fork produces oscillations of the surrounding particles of the air; these, in turn, excite oscillations of particles farther away, and thus the sound wave is formed.

When a simple periodic wave is travelling in a medium, the particles of the medium are oscillating. If the *amplitude* of the wave (the height of the crest for waves on water) is small, the amplitude of the oscillations is proportional to the amplitude of the wave, and the shape of a larger wave is similar to that of a smaller one. For waves of large amplitude, the picture may be quite different. For example, high waves on water are much steeper than low ones. "White caps" may emerge on such waves, and eventually they may overturn. The movements of particles in such a wave are irregular and chaotic. This is an extreme demonstration of the nonlinear nature of a high wave.

To understand *nonlinearity*, one should first understand *linearity*. Characteristic properties of any linear oscillation are: (1) that its period is independent of the amplitude; (2) that the sum of two linear oscillations is also a linear oscillation; and (3) that small amplitude oscillations are linear. Linear waves have similar properties: (1) the shape and velocity of a linear wave are independent of its amplitude; (2) the sum of two linear waves is also a linear wave; and (3) small amplitude waves are linear. Large amplitude waves may become nonlinear, which means that these simple rules do not apply.

Nonlinearity results in the distortion of the shape of large amplitude waves, for example, in the formation of "white caps" on waves, in turbulence, and in many other complex phenomena, some of which will be discussed in this book. However, there is another source of distortion. It is well known that waves having different *wavelengths* (the length from crest to crest) move with different velocities — a phenomenon known as *dispersion* of a wave. A familiar effect of dispersion occurs when observing what happens to a wave when a stone is thrown into the water. Clearly, long waves move faster than short ones. You may further observe that a "hill" on water — a solitary pulse, or, a "piece of wave," so to speak — usually degrades very fast.

Remarkably, some such pulses do not disappear and can travel in water for a long time, preserving their shape and velocity. It is not easy to observe this unusual *solitary wave*, and even more difficult to explain. The first observation of this remarkable phenomenon was made more than 160 years ago, and puzzled scientists for decades. About 100 years ago the mathematical equations describing solitary waves were solved, at which point it was recognized that the solitary wave may exist due to a precise balance between the effects of nonlinearity and dispersion. Nonlinearity tends to make the hill steeper, while dispersion flattens it. The solitary wave lives "between" these two dangerous, destructive "forces."

The most striking properties of the solitary wave, in particular the fact that the wave has properties common to particles, were uncovered and understood after 1967. Indeed, the solitary wave on water is a member of a large family of particle-like waves now generically called *solitons*. In the 20th century, many solitons existing in very different physical media were discovered and studied, but their common features have become clear only in the last three decades.

Solitons are now studied in such diverse sciences as biology, oceanography, meteorology, solid state physics, electronics, elementary particle physics, and cosmology. Solitons certainly exist in crystals (dislocations), in magnetic materials (domain walls), in superconductors and superfluids (vortices), in the atmospheres of the earth (tornados) and of other planets, in the ocean (tsunami), in galaxies (vortices), and in living organisms (nerve pulses). In all likelihood, solitons also played a role in the early history of our universe. Modern theories of the smallest building blocks of matter (superstrings or more complex objects — supermembranes) predict new types of solitons that are, unfortunately, inaccessible to experiment. On the other hand, some solitons are currently used for storing and transmitting information (solitons in optical fibers) or may be used in the near future (solitons in superconductors). Sooner or later you will encounter solitons.

This book is the first attempt to explain the idea of the soliton within the context of its historical development, from first observations to recent applications. Although the reader is assumed to have some knowledge of high school physics and mathematics, the book can be read at three different levels.

No knowledge of mathematics and only an intuitive knowledge of physics is required at the first level. The reader may skip over formulas and detailed physical explanations and enjoy the descriptive history of the development of 19th century science.

To better understand the soliton, however, an effort must be made to intuitively grasp the elementary theory of oscillations and waves. Thus, at the second level, one may still skip over the formulas and concentrate only on the graphical representations of the concepts.

On the third and deepest level, however, the reader must follow elementary mathematical computations and perhaps use the formulas to make experiments on the computer. Mathematical equations that require more advanced mathematics are collected in an appendix. Finally, for those who would like to delve deeply into the real theory of solitons, a reference list is provided to modern scientific literature on the subject.

In Part II of the book, some knowledge of elementary mathematics is required to understand the rather thorough explanation given here of nonlinear oscillations of the pendulum, dispersion of waves, phase and group velocity. Part III, which is perhaps the most interesting, may be read almost independently of Part II.

The overall presentation may appear somewhat irregular at first, but with some patience, the reader will arrive at an understanding of the deep connections between apparently different ideas and diverse phenomena. A primary goal of this book has been to uncover these connections which serve to eventually demonstrate the deep unity of science. The history of the soliton is a generic example of how a significant scientific idea develops. From a simple observation that generates controversial interpretations, an idea can become a widely-accepted mathematical theory that penetrates different areas of the natural sciences, and finally becomes embodied in physical devices that can change our everyday life. It would be much simpler to present this development in a rigid logical (or historical) order, but then the flavor of creative work, full of uncertainties, controversies and misunderstandings, would be completely lost.

Whenever possible and desirable, I try to adhere to a logical and rigorous style of exposition and give all details. Sometimes, however, I rather choose to give an overall impression and leave many details cloudy, or give only abstract sketches.

Scientific events are carefully ordered chronologically, although I move freely in time throughout this presentation. The reason for these "travels in time" is the following: I wanted to show that although the number of scientists, the instruments used, and the role of science in society have all dramatically changed in the last 150 years, the basic methods of scientific thinking have remained the same throughout different epochs.

This book has its own history. First published in 1984 in the former Soviet Union, the second Russian edition was somewhat enlarged in 1990. The present version is neither completely new nor a literal translation of the Russian original. While the "skeleton" is preserved, the "flesh" has been renewed. Several addenda are included and my aim has been to reach a western audience.

Before beginning the story of the soliton, I would like to make some general remarks concerning one of the main motifs of this book, which, although it lies in the background, essentially determines the content, the structure and even the style of this work.

The soliton is a multifaceted scientific phenomenon that is relevant to many scientific fields. A book on this subject could be written by a mathematician, by a specialist in hydrodynamics, by a solid-state physicist, by an expert in radio-engineering and electronics, or by a historian of science. I am none of these but, as a theoretical physicist, I am supposed to be able to apply fundamental physical principles to any of the natural sciences. Unfortunately, in today's world of specialization — I am a specialist in quantum field theory and high energy physics — it is impossible to have enough expertise in all the scientific areas in which the soliton concept is important and must be treated in this book.

However, theoretical physics is based on a few fundamental concepts, such as space, time, particles, waves, and on the laws of physics describing movements of particles and propagation of waves in space, as well as interactions between them (collisions of particles or emitting waves by moving particles, and so on). The laws of physics are expressed by rather abstract mathematical equations. The most difficult thing is to explain how a theoretical physicist handles real physical objects and their abstract concepts. Physics starts with simple observations of events occurring in the real world but, to find mathematical laws describing such events, the physicist has to first give a quantitative description of them. The concept of the point particle having a certain position in space was introduced in this way. By measuring distances between positions of the same particle at different moments of time, we can find the velocity of the particle. Applying the same procedure many times, we can find how the velocity depends on time and thus find the acceleration of the particle.

The first fundamental law of physics is the statement that acceleration is proportional to the force acting on the particle. This is Newton's celebrated second law of mechanics (dynamics). The most difficult thing in this law is the force. Newton knew about the force of gravity and elastic forces. Electric and magnetic forces were discovered later, but this did not change the Newtonian picture of the physical world, which was considered as a collection of moving particles with forces between them. All material bodies (gases, liquids, solids) were regarded as complexes of particles, and all physical phenomena, according to this view, could be described and explained in terms of movements of interacting particles. For example, waves in water were described as resulting from coherent oscillatory movements of small particles of water. Similarly, this applies to sound waves travelling through the air or through solids. The mathematical theory of these waves, based on the Newtonian picture, was finally formulated in the beginning of the 19th century.

At the same time, this consistent picture of the world was shaken up by experiments that proved the wave nature of light. The problem was that light waves may exist even in empty space. Physicists immediately reacted to this problem by inventing the "ether," a substance in which waves of light propagate but do not interact with matter and do not consist of any known particles. The problem of the ether became even more difficult in the Faraday and Maxwell theories of electromagnetic phenomena formed in the second half of the 19th century. The notion of ether had no place as it became clear that light also has an electromagnetic nature. By the end of that century, the concept of ether was finally abandoned by the majority of leading physicists, and Faraday's concept of the electromagnetic field was accepted.

This was "the beginning of the end" of the Newtonian view of the world. The 20th century began with reconsideration of the Newtonian concepts of space and time in the theory of relativity. Even deeper was the quantum revolution that made the physics of the 20th century radically different from the physics of the 19th century.

Although the general approach of physicists to physics problems has not radically changed since the end of the 19th century, modern concepts of space, time, particle, and wave are very different. In quantum theory, there is no contradiction between particle and wave pictures of light phenomena; they are complementary. Moreover, in quantum theory, movements of any particle are described by certain "waves of probability." This very abstract mathematical concept is absolutely necessary for formulating the laws of physics on the microscopic level. But, on the macroscopic level, the laws of 19th century physics are still generally applicable, except for some aspects of low temperature physics.

The soliton observed 160 years ago can be completely understood in terms of the physics and mathematics of that time. However, its most important particle-like properties could not have been discovered in that century; the idea of a "particle-like" wave was absolutely foreign to 19th century physicists. An exact and complete mathematical theory of the soliton could not have been established before the middle of the 20th century. Although the soliton is not a quantum object, its mathematical description requires tools developed in modern quantum physics, and a complete theory of solitons is a rather refined branch of modern mathematics.

I will not try to instruct the reader in the ABCs of this mathematics. Instead, I will first describe how the concept of the soliton gradually emerged from the first observations of it as a solitary wave in water and how this idea became more and more abstract and mathematical, until it eventually penetrated many branches of physics and other natural sciences.

My view of the soliton is that of the theoretical physicist. This means, in particular, that I am always looking at common features of diverse solitons, trying to uncover conceptual simplicity in apparent complexity. I will gradually introduce the most important concepts of theoretical physics and basic principles of a theoretical approach to understanding the soliton's physical reality.

Many people believe that theoretical physics is a very difficult science. A great German mathematician David Hilbert (1862–1943), who made many important contributions to mathematics, including the mathematical apparatus of quantum physics and of general relativity, once said that physics is much too hard for physicists. He obviously had in mind quantum theory, and I agree with this statement. However, theoretical physics is easier to understand than any other natural science. It deals with extremely simplified mathematical models of real phenomena and uses only a few basic concepts. The main concepts and results of theoretical physics as related to solitons can be explained without using much mathematics. Of course, some physics intuition and acquaintance with basic mathematics is desirable to follow my presentation of solitonic phenomena.

I recommend the book written by one of the creators of modern theory of superconductivity, Leon N. Cooper, *An Introduction to the Meaning and Structure of Physics*, Harper and Row, Publ. NY, 1968. The evolution of the main concepts of theoretical physics – space, time, particle, wave, field, quantum – is nicely presented in the book by Albert Einstein and Leopold Infeld, *The Evolution of Physics*, Simon and Schuster, N.Y. 1938.

Acknowledgments

When preparing the first (Russian) version of this book I used many scientific and popular science books, as well as original and review papers. I mentioned some papers and their authors in the text, but it was impossible to mention and to thank everyone. Especially useful were my contacts with S.P. Novikov, Ya.A. Smorodinskii, and L.G. Aslamazov who made many critical and very useful remarks. The second Russian, and especially, the English version of the book were also strongly influenced by many people. I have to mention in particular Ya.B. Zeldovich, N. Zabusky, Ch. Zabusky, and E. Beschler. To all these people, I am really very grateful.

The Versatile Soliton

Part I

An Early History of the Soliton

And science dawns though late upon the earth...

Percy Bysshe Shelley (1792–1822)

To understand a science it is necessary to know its history

Auguste Comte (1798–1857)
Positive Philosophy

The first officially registered scientific encounter with what is now called the soliton happened more than a century and a half ago in August of 1834 near Edinburgh, Scotland. Scottish scientist and engineer John Scott Russell (1808–1882) was well prepared to make this first observation and gave an accurate, clear, and even poetic account of his first meeting with the soliton, which he called the *translation wave* or the *great solitary wave*.

Although Russell devoted many years to further investigation of this phenomenon, his colleagues neither saw its significance nor shared his enthusiasm. Between 1844, when Russell's paper was published, and 1965, fewer than two dozen papers relating to the solitary wave were published. In 1965, American physicists Martin Kruskal and Norman Zabusky, who solved computationally an important equation related to this phenomenon, coined the term *soliton*.

Now, thousands of scientific papers are published every year in which the role of solitons in physics, mathematics, hydromechanics, astrophysics, meteorology, oceanography, and biology are studied. International scientific conferences devoted to solitons and related problems are organized all over the world. More and more scientists join the "society" of soliton hunters. In fact, the concept of the soliton and its technical applications are becoming so important that the time has come for every educated person to become acquainted with the phenomenon, which is the *raison d'être* for this book.

Because the soliton's discovery, its delayed acceptance, and its relationship with technology are inextricably bound in developments that occurred through the history of science, this book will provide a framework for understanding the soliton, shining a light on those scientists whose work contributed to its "rise." So let us start from the beginning and, to better understand the reaction of Russell's colleagues to his discovery and ideas, try to imagine that we are in the first half of the 19th century, around the year 1834.

Chapter 1
A Century and a Half Ago

The Nineteenth Century, the Iron
And really hardhearted Age ...

Aleksandr Blok (1880–1921)

Poor our Century! How many attacks on him, what a monster is he said to
be! And all this for the railways, for the steamships — his great victories not
only over matter but over space and time.

Vissarion Grigorievich Belinskii (1810–1842)

Only five years prior to 1834 the first railroad had been completed, and steamships were just beginning to be built. In industry, steam engines were the most
popular sources of power. In fact, this period is sometimes called the "Steam
Revolution." It was also the time of Napoleonic wars, social disorders and revolutions, of Goethe, Beethoven, Byron and Pushkin, and of many important scientific
discoveries, including many principal discoveries in physics, the full meaning of
which would become clear only gradually many years later. At the time, physics
was not yet regarded as a separate science. The first physics institute was created
in 1850 in Vienna. Around 1834, university professors were still reading courses
in "natural philosophy," which included what we call experimental, theoretical
and applied physics.

Because we are privileged now to know the role that the new concepts in
physics played in the development of western civilization, it is not easy for us
to imagine what educated people in the 19th century might have thought of the
science of their time. In addition, the role of science in the 19th century was very

A.T. Filippov, *The Versatile Soliton*, Modern Birkhäuser Classics,
DOI 10.1007/978-0-8176-4974-6_1, © Springer Science+Business Media, LLC 2010

different from what it would become in the second half of the 20th century; its practice was also different.

Those were the "golden" years of physics, although no one then could have imagined our supercomputers or superconducting supercolliders of elementary particles! Using extremely simple tools, however, some fundamental discoveries were made. One of the most amazing discoveries of the first part of the 19th century was made using a small piece of ordinary wire with electric current and a standard compass (magnetic arrow). Hans Christian Oersted (1777–1851) was extremely attracted to the idea of mutual interrelations among such diverse natural phenomena, as heat, sound, electricity, and magnetism. So it happens that in 1820 at Copenhagen University, while giving a lecture devoted to searching for a connection between magnetism, "galvanism" and "inorganic" electricity,[1] Oersted observed a striking phenomenon. When the electric current was allowed to travel through the wire, which was parallel to a magnetic arrow, the arrow changed direction! One of his pupils later remarked that "he was quite struck and perplexed to see the needle making great oscillations." Oersted was so astounded because the original aim of the experiment was to demonstrate the "well-known" absence of a relationship between magnetism and electricity produced by the Volta battery; apparently, nobody had ever tried to close the circuit.

Oersted's observation was immediately followed by a series of experiments clarifying quantitative details of the relationship between electricity and magnetism. The investigation of this relationship triggered an avalanche of exciting scientific developments, in which the key figure was André Marie Ampère (1775–1836). In his famous series of papers published between 1820–1825, Ampère gave the mathematical expression for the magnetic force produced by the electric current (Ampère's law) and, more generally, laid the foundation of a unified theory of electricity and magnetism which he called *electrodynamics*. In turn, Ampère's work laid the groundwork for the chain of great discoveries made by the self-taught genius, Michael Faraday.

Faraday began publishing the results of some of his experiments in 1821, but he made his main discoveries during the decade 1830–1840. In 1831 he first observed the electromagnetic induction ("electric current without batteries"), and later formulated the quantitative "Faraday's law" describing the electric current induced by a changing magnetic field. It is interesting to note that Ampère had stumbled upon this phenomenon in his experiments in 1822, but did not pay any attention to it. On the contrary, Faraday, having some hypotheses about the relationship, conducted many experiments before he finally succeeded in observing induction. He also immediately realized the possible practical value of his discov-

---

[1] "Galvanism" was so named after the Italian scientist Luigi Galvani (1737–1798), who discovered in 1780 electric phenomena in living creatures, typically, frogs. "Inorganic" electricity was known to ancient Greeks, but practical applications of this "type" of electricity began only after 1792 with discoveries by another Italian scientist Alessandro Volta (1745–1827). Although the ancient Greeks thought that electric and magnetic forces were of the same origin, the common belief before Oersted's discovery was that they were independent phenomena.

ery, and in 1832 he constructed the first generator of electric current. Continuing to alternate between performing imaginative, beautiful experiments and making deep, unorthodox reflections about them, in 1852 Faraday came to develop the concept of the *electromagnetic field*. In his experiments, Faraday used equipment which was not much more complicated than Oersted's.

In 1853 Hermann Helmholtz, the esteemed German scientist, wrote of a visit to Faraday's laboratory:

> I have succeeded in making the acquaintance of Faraday, who is the leading physicist of England and Europe.... He is artless, kind, and unpretentious like a child; I never met a more attractive man.... He was invariably obliging and showed me everything worth seeing. But it was not much to look at, because he is using for making his great discoveries old pieces of wood, wire, and iron.

Maxwell will later write:

> It is to be hoped that the nobleness of his simple, undramatic life will live as long in men's memories as the discoveries which have immortalized his name. Here was no hunger after popular applause, no jealousy of other men's work, no swerving from the well-loved, self-imposed task of "working, finishing, publishing."

A 20th century English novelist Aldous Huxley (1894–1963) would say: "Even if I could be Shakespeare I think I should still choose to be Faraday."

Let us recall that the electron had not yet been discovered. The idea that an elementary electric charge must exist suggested itself to Faraday in 1834, when he discovered the laws of electrolysis, and they seemed easier to explain using this concept. Despite this, however, the electron became scientifically established only at the end of the 19th century. The term "electron" was not introduced until 1891.

In 1834, in spite of all the successes in the studies of electricity and magnetism, a mathematical theory describing electromagnetic phenomena was far from being established. In fact, the father of this theory, James Clerk Maxwell, was at that time three-years-old and living with his family in Edinburgh, the same city where Russell was giving lectures in natural philosophy. As this field encompassed various sciences, I suspect that Russell devoted little time to electrodynamics and, if he mentioned it at all, he would hardly have used any mathematics in his explanation. At that time, electrodynamics was not a mathematical theory. Although Maxwell would say later that he was building not only upon Faraday's ideas, but also his mathematical methods, one must not accept this statement literally because Faraday did not use even elementary algebra in his work. The real meaning is that Maxwell succeeded in translating Faraday's ideas into mathematical language. In his famous *Treatise on Electricity and Magnetism* (1873) he explained the relation of his work to Faraday's ideas very clearly:

It was perhaps to the advantage of science that Faraday, though thoroughly conscious of the fundamental forms of space, time and force, was not a professional mathematician. He was not tempted to enter into the many interesting researches in pure mathematics which his discoveries would have suggested if they had been exhibited in a mathematical form, and he did not feel called upon either to force his results into a shape acceptable to the mathematical taste of the time, or to express them in a form which mathematicians might attack. He was thus left at leisure to do his proper work, to coordinate his ideas with his facts, and to express them in natural, untechnical language.

... before I began the study of electricity I resolved to read no mathematics on the subject till I had first read through Faraday's *Experimental Research in Electricity*. I was aware that there was supposed to be a difference between Faraday's way of conceiving phenomena and that of mathematicians, so that neither he nor they were satisfied with each other's language. I had also the conviction that this discrepancy did not arise from either party being wrong...

As I proceeded with the study of Faraday, I perceived that his method of conceiving the phenomena was also a mathematical one, though not exhibited in the conventional form of mathematical symbols. I also found that those methods were capable of being expressed in the ordinary mathematical forms, and thus compared with those of the professional mathematicians.

For instance, Faraday, in his mind's eye, saw lines of force traversing all space where the mathematicians saw centres of force attracting at a distance: Faraday saw a medium where they saw nothing but distance: Faraday sought the seat of the phenomena in real actions going on in the medium, they were satisfied that they had found it in a power of action at a distance impressed on the electric fluids.

In contrast to electrodynamics, the mechanics of point-like particles and of solid bodies, as well as the theory of the movement of fluids (*hydrodynamics*), were essentially already mathematical theories. Thus, all the problems of the dynamics of point masses (point-like particles) were essentially reduced to precisely formulated mathematical problems using the methods of *ordinary differential equations*. Examples of this are Newton's equations (1687) and the more general Lagrange equation (1788). Similarly, the problems of hydromechanics were mathematically formulated with the aid of *partial differential equations*, examples of which are Euler's equations describing movements of fluids (1755).[2] The equations most appropriate for describing motions of real water were proposed in 1823 by a French scientist, Louis Marie Henri Navier (1785–1836)[3] of *L'Ecole Polytechnique*. In summary, by this time, mechanics and hydromechanics had reached the age of maturity. This means that all the physical principles were clearly formulated and written in mathematical form.

---

[2]These were later to be rewritten in a somewhat different form by Lagrange.

[3]Of course, other scientists also contributed to formulating the final equations, e.g., D. Bernoulli, but in giving an overview, I am not going into all the details.

Thus, it was quite natural that these well-developed branches of science were considered a foundation for constructing theories involving recent discoveries in physics; for any 19th century physicist, *to understand a new phenomenon was to explain it in terms of mechanical laws*. In this sense, *celestial mechanics* was considered a paradigm of scientific theory. The most comprehensive presentation of this science was given by Marquis Pierre Simon de Laplace (1749–1827) in his monumental work *Treatise on Celestial Mechanics*, five volumes of which were published in the first quarter of the 19th century. In this book, Laplace collected and critically expounded upon the results of the giants of the era in the area of mechanics: Bernoulli, Euler, D'Alembert, Lagrange, and himself. The influence of this book on the formation of the mechanical sciences and of a "mechanical view of the world"[4] was crucial, and for many scientists of that time it was a virtual bible.

By the way, in 1834, the Irish mathematician William Rowan Hamilton (1805–1865) added a final stroke to the beautiful picture of Newton–Lagrange mechanics. He rewrote the Lagrange equation in the so-called "canonical" form and discovered an analogy between optics and mechanics. Hamilton's canonical equations later played a very important role in mathematical formulation of the laws of statistical mechanics (L. Boltzmann and J. W. Gibbs). His "optical-mechanical" analogy, establishing a relation between wave propagation and particle movements, was used in the 20th century in finding the basic laws of quantum physics (quantum mechanics). His general analysis of the relationship between particles and waves played a role in the development of soliton theory.

While in the 19th century electrodynamics was a pure science, mechanics and hydrodynamics, as well as the theory of deformation of solids (*elasticity theory*), were to a large extent applied sciences, the development of which was strongly stimulated by the needs of the fast developing field of engineering. All in all, physical concepts and mathematical models related to the mechanical sciences were dominant in mathematical physics until the end of that century.

Even Maxwell, creator of the electromagnetic theory, which finally toppled mechanics from its seat of dominance, was strongly influenced by this way of thinking, his prior research having included elasticity theory, applied theory of mechanical stability (used in designing regulators of steam engines, etc.), and other branches of applied mechanics. Consequently, it must have seemed quite natural to Maxwell that the equations of electromagnetic theory should be very similar to the equations of hydrodynamics and elasticity theory. Moreover, it is documented that he hoped that a final theory of electromagnetic phenomena could be given a mechanical explanation, i.e., in terms of movements and deformations of a medium known at the time as "ether."

In forming these views, Maxwell was influenced by another famous Scottish physicist, William Thomson (1824–1907), later called Lord Kelvin in recognition

---

[4] According to Laplace, all natural phenomena could be (at least, in principle) described in terms of the basic laws of mechanics. This view dominated scientific philosophy of the 19th century.

of his scientific achievements and service to society. Like Laplace, Kelvin was such a firm proponent of the mechanical view of the world that he believed all natural phenomena could be described and eventually explained by the laws of mechanics. He himself expressed this very clearly: "I am never content until I have constructed a mechanical model of the subject I am studying. If I succeed in making one, I understand; otherwise I don't." This conviction led him to oppose Maxwell's theory, and only by 1899, after Lebedev's experiments demonstrated the pressure of light, had he come to recognize its truth. Most other leading physicists were convinced much earlier.

Beginning of Wave Theory

> Nature herself takes us by the hand and leads us along by easy steps. . . . The problem well put is half solved.

> Josiah Willard Gibbs (1839–1903)

In the 1820s, although the main equations describing the motions of liquids had been established, the mathematical theory of waves in water was at a rudimentary stage. An elementary theory of waves on the surface of water was suggested in Newton's *Philosophiae Naturalis Principia Mathematica*, first published in 1687. A century later the French-Italian mathematician Joseph Louis Lagrange (1736–1813), who called this book "the greatest creation of the human mind," found that Newton's theory of waves on water was based on the incorrect assumption that the particles of water in the wave harmonically oscillate in a vertical direction. Nevertheless, Newton's problem was correctly formulated, and Lagrange used it to find a more accurate approach to water waves. He constructed a new theory for two simple cases: first, for the *small amplitude* waves and, second, for the waves in reservoirs having small depth compared to wavelength (*shallow water* or *tidal* waves). Being occupied with more general and deeper mathematical problems, he did not develop the theory in full. The complete mathematical theory was developed by other scientists.

> Will you find many individuals who, while admiring the beautiful play of waves in a brook, are thinking about equations mathematically describing the shapes of the waves?

> Ludwig Boltzmann (1844–1906)

Soon after Lagrange published his work, an amazingly simple and mathematically exact solution of the hydrodynamic equations for waves was derived. In 1802, a Czech scientist František Jozef Gerstner (1756–1832) obtained the first exact solution describing water waves of arbitrary amplitude. These *Gerstner waves* may exist on *deep water*. This means that the water depth (with no waves on the surface) is large compared to the wavelength.

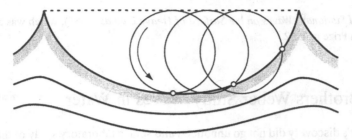

Figure 1.1 Gerstner's wave. The upper curve, formed by a uniformly
moving circles, is a cycloid; curves below are trochoids

The simplicity of the Gerstner wave is based on a surprising fact — the droplets
of water in this wave move in circles— as illustrated in Figure 1.1.

It is easy to observe the following simple properties of the Gerstner wave. A
water droplet at the surface moves uniformly in a circle of radius equal to the amplitude of the wave. The period of rotation is equal to the period of wave motion.
The droplets have maximum velocities in the trough (backward with respect to
the wave propagation) and on the crest (in the forward direction). Less obvious
is the fact that the droplets below the surface move in smaller circles and that the
circle's radius decreases quickly with increasing depth. Several wavelengths below the surface the radius is negligibly small and so the droplets are practically
resting.

It is obvious that the shape of the Gerstner wave is not sinusoidal. Actually, it is
cycloidal — from the Greek word $\kappa\upsilon\kappa\lambda\omega\sigma$, circle, and $\epsilon\iota\delta\omega\sigma$, shape. The cycloid
is the curve traced by a point on a wheel moving on a plain road. This curve is also
called the *trochoid*, from the Greek word for wheel, $\theta\rho\omega\xi\omega\sigma$, and the Gerstner
wave is often called the *trochoidal* wave.[5] The Gerstner wave becomes close to
sinusoidal when its amplitude is very small, but even in this case the water droplets
move in circles, and Newton's assumption is not valid.

In fact, Newton knew that the water droplets move in circles and used his simplifying assumption only for estimating the velocity of waves in water. He defined
the velocity of the wave as $v = \lambda / T$ where λ is the wavelength. Then he showed
that v is proportional to $\sqrt{\lambda}$. We will see below that this result is correct and find
the proportionality factor which was known to Newton only approximately.

As far as I know, the exact Newton-like theory of waves with the circular motion of the
droplets was first given in 1863 by a Scottish engineer and physicist, William John Rankine (1820–1872), Professor at the University of Glasgow. The analytical theory of periodic
waves originated by Gerstner was completed in the 20th century in the work of Aleksandr
Ivanovich Nekrasov (1883–1957) a professor at Moscow University. The precomputer
period in the theory of large amplitude waves was summarized in his basic work *Exact*

---

[5]Note that the points on the rim of the wheel trace the cycloid. The trochoids are formed by the
paths of the points between the rim and the center of the wheel.

Theory of Stationary Waves on the Surface of Heavy Liquids (1951), which was awarded
the Stalin Prize in 1952.

The Brothers Weber Study Waves in Water

Gerstner's discovery did not go unnoticed and soon a laboratory study of the water
waves was undertaken by the Weber brothers.

The eldest, Ernest Weber (1795–1878), made important discoveries in anatomy
and physiology, especially in neural physiology. In 1834 he published a paper that
opened the way to a quantitative and even mathematical approach to neural physiology. Wilhelm Weber (1804–1891) became a physicist. He worked for many
years with K. Gauss, one of the greatest mathematicians of the 19th century (often called by his contemporaries the "king of mathematicians"). Supported by
Gauss, he organized the first laboratory specializing exclusively in physics at the
University of Göttingen in 1831. W. Weber is best known for his work in electricity and magnetism. In fact, before Maxwell's ideas were accepted, Weber's
theory of electromagnetic phenomena was popular. He was the first to introduce
the concept of "particles of electricity" (predecessors of elementary charges). He
also proposed the first "planetary-like" model of the atom in 1846, developed in
detail one of Faraday's ideas about magnetics, and invented several useful physics
instruments, e.g., magnetometers for precise measurements of the magnetic field
of the Earth.

Around 1823, Ernest, Wilhelm and the youngest brother Eduard decided to
study waves. They constructed a very simple but effective apparatus which is
called *Weber's tray* (its modern descendant, a wave trunk, is still used by exper-
imentalists). This is simply a long box with one glass wall and with a device for
exciting water waves. Using this instrument they performed many experiments
with waves; in particular, they had checked the Gerstner theory by investigating
Gerstner waves.

The Webers published the results in a book *Die Wellenlehre* (Wave Science) in
1825. This historical first treatise on an experimental study of waves presented a
detailed consideration of waves of different shapes and the relationship between
their velocities, wavelengths, amplitudes, etc. The experiments were conceptually
simple but ingenious and effective. For instance, to study shapes of waves, they
put into the tray a plate made of mat glass. When the wave reached the middle
of the tray, they pulled the glass out quickly so that the front part of the wave
was accurately imprinted on the plate. To watch the paths of the water droplets in
the waves, they filled the tray with muddy water from the Saale river. Observing
the waves with a weak microscope, they found the shapes of the trajectories. In
this way they proved that the trajectories are circular near the surface and become
elliptical closer to the bottom, and that the droplets move horizontally on the bottom. Through such observations the Weber brothers discovered many interesting
properties of waves in water and in other liquids as well.

From Waves in Water to Waves in Ether

What is now proved was once only imagined.

William Blake (1751–1827)

The aims of scientific thought are to see the general in the particular and the eternal in the transitory.

Alfred North Whitehead (1861–1947)

Meanwhile Lagrange's ideas were being independently developed by mathematicians, especially by Augustin Louis Cauchy (1789–1857) and Simon Denis Poisson (1781–1840). A contribution to this development was also made by a compatriot of mine, Mikhail Vassilievich Ostrogradskii (1801–1862). Although all of these scientists are well known and associated with theorems, equations, and formulas, their investigations into the mathematical theory of water waves are not so widely known. In particular, they all considered small amplitude surface waves which are encountered in sea storms, behind moving vessels, on sandbanks, etc. It is evident that the theory of such waves is very useful in solving important engineering problems related to sea waves.[6] It was not easy to foresee applications of this theory to other branches of pure and applied sciences that apparently have nothing in common with hydrodynamics. The development of mathematical ideas and methods has its own logic, and usually it is difficult to judge whether a mathematical theory or a solution to a particular problem is "practically" useful. It often happens that seemingly "useless" concepts or solutions to abstract problems of purest mathematics later prove to be of great value for physics or even technical applications.

The same can be said about deep physical concepts, for example, Faraday's hypothesis concerning the wave nature of propagation of electric and magnetic forces in space. Faraday was recognized as one of the world's greatest physicists even during his own lifetime. Nevertheless, it is not widely known that he was seriously interested in water waves. As we know, he was not a master in standard mathematical methods, and certainly not the rather complex methods of the wave theory developed by Gerstner, Cauchy, Poisson, Ostrogradskii, and others. However, he had a very deep knowledge of wave properties and a very clear intuitive comprehension of wave concepts.[7] In his attempts to explain how electric and magnetic fields propagate in space, he used the analogy with water waves.

---

[6] The importance of the theory of waves was recognized by the Paris Academy of Science which in 1815 suggested a "Grand Prize in Mathematics" for a contribution to water wave theory. The winner was Cauchy and the runner-up was Poisson. Ostrogradskii had written his work on waves in 1822 while imprisoned in a jail for debtors in Paris in Clichy and then sent it to Cauchy. Cauchy was so impressed that he paid Ostrogradskii's debts, although he was not a rich man himself.

[7] It is interesting that his intuitive understanding could not be adequately expressed in terms of the mathematics of his time. In fact, his ideas about the electromagnetic field were adequately translated into mathematical language only in the 20th century.

It was possibly this analogy that finally led him to the revolutionary hypothesis that there must exist waves of electric and magnetic fields which propagate in space with a finite velocity. This hypothesis was certainly ahead of its time and was difficult to prove. Probably for these reasons, Faraday didn't publish these ideas at the time, but he described them in a letter which he placed in an envelope, sealed and deposited in the archives of the Royal Society on the twelfth of March, 1832. The fate of this letter was unusual. Faraday forgot about its existence (possibly because of a developing illness which resulted in a partial loss of memory), and it was not discovered in the archives until 1938! Fortunately, however, Faraday had meanwhile described his ideas about electromagnetic waves in a paper published in 1846, where he also advanced the hypothesis of the electromagnetic nature of waves of light. Note that in 1845 Gauss also came to the conclusion that electric and magnetic interactions propagate with a finite velocity.

Of course we cannot reconstruct the chain of thoughts that led Faraday to his remarkable hypothesis. However, his experiments with water waves, as well as some of his reflections on waves in general, may be found in a paper published shortly before he wrote the famous letter. In it he investigated ripples on the surface of water, i.e., waves of very small amplitude. Physicists call them surface tension waves, as their existence is due to the forces of surface tension[8] and has nothing to do with gravity forces responsible for the usual water waves described above. To study these surface tension waves, Faraday invented, with his usual ingenuity, a very simple apparatus. Russell knew Faraday's paper and used the same apparatus for further experiments with capillary waves. These experiments of Faraday and Russell are described in a nicely written book by Rayleigh, *The Theory of Sound*, first published in 1877 and still worthwhile.

Scientific Discoveries — Mainstream and Ahead of Time

What really makes science grow is new ideas, including false ideas.

Karl Raimund Popper (1894–1963)

Nineteenth century science discovered the technique of discovery...

Aldous Huxley (1894–1963)

We can see that the first half of the 19th century was a very interesting time for science. All the principal laws of mechanics were formulated and were given the form we use today. By mid-century, the basic equations of hydrodynamics

---

[8]These forces cause the water to rise in thin (capillary, from the Latin *capillus*, hair) tubes above the surrounding liquid. The bore of the tube must be small, usually approximately $0.5mm$. Surface tension waves may thus be called capillary waves. We will discuss them in more detail in Chapter 5.

and elasticity theory had been discovered, and many scientists were involved in solving as well as in applying them to practical engineering problems. However, physicists were, at that time, more enthusiastic about light waves.

In the first quarter of the century, English physicist and physician Thomas Young (1773–1829), French physicist Augustin Jean Fresnel (1788–1827), and French physicist, astronomer and politician Dominic François Arago (1786–1853) succeeded in convincing other physicists that light is produced by waves, but it was not an easy task. Many well known and influential scientists thought that light was produced by particles (note that this point of view had been advanced by Newton, although one can also find in his publications some elements of wave considerations). For example, among opponents of the wave theory were Laplace and Poisson. Wave theory was generally recognized only after a dramatic meeting of the Commission of the Paris Academy that was devoted to considering Fresnel's paper, presented to the Academy competition in 1819. In the Commission report we read:

> One of the Commission members, Mr. Poisson, deduced from the integrals presented by the author the following strange result. The center of the shadow from a large opaque screen must be illuminated exactly in the same way as in the case when the screen is removed.... This prediction was checked in a direct experiment which completely confirmed the presented derivations.

The experiment was designed and performed at the Commission meeting by Arago. By the way, Fresnel himself overlooked this unusual consequence of his formulae.

As mentioned above, the following year the Society was excited by Oersted's dramatic discovery. His paper was published in Copenhagen at the end of July. In early September Arago reported this discovery in Paris. From the end of September Ampère was reporting his new results every week (!) at the Paris Academy meetings. In October the Biot– Savart–Laplace law, giving the formal mathematical expression for the magnetic field generated by electric currents, was discovered. The results obtained in this pre-Faraday period were summarized by Ampère in his *Theory of Electrodynamic Phenomena Based Exclusively on Experiments*.

Note how fast sensational scientific news rushed across Europe, without the aid of radio, telephone or even telegraph. The news about Faraday's spectacular discoveries spread equally fast, but it was quite a different story with his theoretical ideas. They were met without any enthusiasm, and so it happened that his beautiful new concept of the electromagnetic field remained foreign for a long time to the scientific community.

As we mentioned earlier, Maxwell was the first man capable of understanding Faraday's ideas and who found a mathematical embodiment for them in terms of contemporary mathematical language. But this happened later, in the mid-century. One may wonder why Faraday's ideas were more difficult than Ampère's. The answer probably is that Ampère's electrodynamics was closer to the spirit of his age. He did not try to introduce new concepts to explain new discoveries. Remaining entirely in the framework of the commonly accepted Newtonian view of the physical world, he simply looked for a mathematical description of the new phenomena. Faraday's ideas about the electromagnetic field were much deeper than Ampère's and truly revolutionary. His work was a unique combination of original experiments and original thinking. It is fate that Maxwell was born at the right time and right place. Otherwise Faraday's ideas might have been ignored for many years.

In his *Treatise* Maxwell returns several times to a discussion of the difference between Faraday's and Ampère's approaches to electromagnetic phenomena. He notes that they even presented their results in a remarkably different style:

> The experimental investigation by which Ampère established the laws of mechanical action between electric currents is one of the most brilliant achievements in science.... The method of Ampère, however, though cast into an inductive form, does not allow us to trace the formation of the ideas which guided it. We can scarcely believe that Ampère really discovered the law of action by means of the experiments which he describes. We are led to suspect... that he discovered the law by some process which he has not shewn us, and when he had afterwards built up a perfect demonstration he removed all traces of the scaffolding by which he had raised it.
>
> Faraday, on the other hand, shews us his unsuccessful as well as his successful experiments, and his crude ideas as well as his developed ones, and the reader, however inferior to him in inductive power, feels sympathy even more than admiration, and is tempted to believe that, if he had the opportunity, he too would be a discoverer. Every student should therefore read Ampère's research as a splendid example of scientific style in the statement of a discovery, but he should also study Faraday for the cultivation of a scientific spirit...

After optics and electromagnetism, the third most important development in physics during the first half of the 19th century was the beginning of the theory of heat phenomena. The first steps originated from the technical problems of steam engines, but general ideas developed slowly and with great difficulties. The now famous paper of Nicholas Leonard Sadi Carnot (1796–1832) *Reflexions sur la Puissance Motrice du Feu et sur les Machines Propres à Développer cette Puissance*, published in 1824, went completely unnoticed for ten years. Its importance was recognized only after the paper of Clapeyron,[9] *Mémoire sur la Puissance Motrice de la Chaleur* (1834), while the unified approach to all heat phenomena (*thermodynamics*) was worked out only in the second half of the century.

The further development of thermodynamics was not easy, as it required a significant confrontation with the views of senior scientists who occupied the principal positions in the universities and in scientific journals. It is no wonder that the major contributions were made by young men who were not even professionals in physics. In 1841 the German physician Julius Robert von Mayer (1814–1878)

---

[9] Benoit Paul Emile Clapeyron (1799–1864), a French physicist and mining engineer, worked from 1820 to 1830 in Russia. In the above mentioned paper he also introduced the so-called Clapeyron–Clausius equation.

calculated the "mechanical equivalent of heat" and analyzed energy transfers in nature. His paper was rejected by a physics journal. His next paper, developing the same ideas in a more complete and clear form, was published in a medical journal. In 1845 Mayer applied the same considerations to living creatures. This profound thinker was unhappy at being regarded by his fellow citizens and even by relatives as insane. In fact, he tried to commit suicide and then spent several years in an asylum. Eventual recognition of his work came too late to restore his broken life. A young Englishman James Prescott Joule (1818–1889) was more fortunate. Having substantial property, he could afford the luxury of performing experiments demonstrating the origin of heat from mechanical work, electric current, etc.[10] In 1845 he published the results of his extensive measurements of the mechanical equivalent of heat. His work was well accepted in Britain and he was elected to be a Fellow of the Royal Society in 1850. Later his work was also recognized on the continent.

Two more epoch making papers should be mentioned. One was published in 1822 by the French mathematician, physicist and Egyptologist[11] *Baron* Jean Baptiste Joseph Fourier (1768–1830). It was called "The Analytical Theory of Heat" and dealt with heat propagation. In this paper Fourier also suggested, developed in great detail, and applied to physics problems a method of expanding arbitrary functions in sinusoidal components. It is now known as the *Fourier expansion* and is applied in practically all branches of science and technology. This Fourier expansion, or Fourier analysis, gradually became one of the main tools for investigating diverse wave phenomena — storms in the ocean, sound, light, radio, etc. The essence of Fourier analysis is that complex wave signals are represented as sums (superpositions) of simple *sinusoidal* (*harmonic*) waves, often called *harmonics* (from *harmony* in music). We will consider harmonics later in more detail.

The second important paper was presented in 1847 by a young German physician, Hermann Ludwig Ferdinand von Helmholtz (1821–1894) at a meeting of the Berlin Physical Society that he had himself founded not long before. He was well known as a physiologist and physician[12] but, from his youth, he was also very much interested in physics, and not only in problems related to biology and medicine. He became famous for his contributions to physics and mathematics. He is justly regarded as one of the greatest scientists of the 19th century and some historians of science consider his paper as one of the cornerstones of the natural sciences.

---

[10]Heat produced by electric current was independently investigated somewhat later by Emilii Christianovich Lenz (1804–1865), known for his work on electromagnetic induction – the *Lenz rule* (1834).

[11]Fourier was also a mentor of young Jean-François Champollion in his early work that eventually led him to decipher hieroglyphic writing. Note that Thomas Young was also an enthusiast of deciphering and devoted much of his time to it.

[12]Through many of his pupils he influenced medical services in many countries including Russia. I first heard his name in my own childhood, when my injured eye was cured at the Helmholtz Institute of Eye Diseases in Moscow.

The title of the paper was "On the Conservation of Force." It was concerned with a general formulation of the *energy conservation law* (at that time energy was called "force"). He considered all known physical phenomena, not only mechanical but also thermal, electric, magnetic, and even physical processes in living organisms (to which he generally referred as "organized creatures"). Applied to heat phenomena, the energy conservation law is now known as the first law of thermodynamics. It was hotly criticized before it was realized that heat is "a degraded form of energy." This idea was mathematically expressed by the second law of thermodynamics, introduced in 1850 by the young German physicist Rudolph Julius Emanuel Clausius (1822–1888). Later, in 1889, J.W. Gibbs would write about this paper of Clausius: "This memoir marks an epoch in the history of physics... the science of thermodynamics came into existence."

Among other results of Helmholtz's work we may mention the observations of the oscillations occurring in the discharge of the "Leiden jar" (electric capacitor). The oscillatory character of the capacitor discharge had already been discovered by the American scientist Joseph Henry (1797–1878), but Helmholtz was apparently unaware of this fact and he rediscovered the oscillations. In addition, he wrote an equation describing these oscillations in mathematical terms. Using his equation, Kelvin soon derived a formula for the period of the electromagnetic oscillations in the closed LC-circuit (the electric circuit containing inductance and capacitance), which we will use later.

Both papers mentioned above were purely theoretical. In fact, they are among the most important theoretical papers of the first half of the 19th century and constitute the foundation of the theoretical approach to physics. Theoretical physics as a science was firmly established and recognized only at the end of the century, after the fundamental theoretical books written by Maxwell, Boltzmann and Gibbs.[13] In the first half century, a purely theoretical approach to physics phenomena was alien to most physicists. The physics community regarded the subject as essentially an experimental science. The most common expressions in the titles of physics papers were "experimental," "based on experiments," "derived from experiments," etc. In view of this, it is no wonder that Helmholtz's paper, which is rightly regarded by historians of science as a model presentation of new concepts and results, was not accepted by a physics journal because it was theoretical and too long. Helmholtz published it later as a monograph. Not long before his death, Helmholtz reflected on his famous paper:

> Young men prefer to undertake solving the most difficult and deep problems. I was no exception, and became absorbed with the mystery of the "vital force".... I have found that... the theory of this force... ascribes to every living body properties of a *perpetuum mobile*.... Reviewing the work of Daniel Bernoulli, D'Alembert and other mathematicians of the last century... I came

---

[13] J.C. Maxwell, *Treatise on Electricity and Magnetism* (1873); L. Boltzmann, *Lectures on Gas Theory* (1896); J.W. Gibbs, *Elementary Principles in Statistical Mechanics* (1902). These three books are highly recommended for an understanding of the origin of theoretical thinking in physics.

across the following question: what relationship must exist between different forces of Nature if the *perpetuum mobile* is impossible and are these relations satisfied in reality.... My first intention was only to give a systematizing and critical discussion of the main facts for physiologists. I would be not surprised if some experts will tell me, "Why, these things are well known! What does this young medical man want, talking in such detail about these matters?" To my astonishment, the authorities in physics whom I happened to contact on these matters were of a quite different opinion. They were rather inclined to reject the validity of the law... and my work was regarded as a fantastic intellectual exercise. Only Jacobi, the mathematician, recognized the relationship of my reasoning with some thoughts of mathematicians of the last century. He became interested in my work and defended me from misunderstanding.

This quotation gives a rather clear and vivid picture of the intellectual climate of that time. But the reaction of the scientific establishment to new ideas is normally similar to this, which is not only natural but, in some sense, even necessary! So let us not blame Laplace for his resistance to Fresnel's ideas, or Weber who thought that Faraday's ideas were wrong, or Kelvin, who for many years resisted Maxwell's theory. Let us rather ask ourselves whether it is easy for any of us to understand radically new ideas, ideas that are totally foreign to all that we know. Let us rather admit that we are all somewhat conservative by nature, and scientists are no exception. It is even said that some healthy conservatism is necessary to prevent science from groundless fantasies. True as this may be, it was no consolation for the geniuses whose lives were broken because their contemporaries did not understand them. As the Soviet writer Mikhail Zoschenko once said, "Our history is our common sin. We all must be ashamed of it."

The most dramatic examples of such conflicts in the first half of the 19th century can be found in mathematics. At that time, the central figures in mathematics were undoubtedly Gauss and Cauchy, who were leaders in completing the great structure of Differential and Integral Calculus, the mathematical base of all modern science. Their work was widely respected and regarded as the most important accomplishment of the time. Unfortunately, those who were at that time looking for new ideas out of the mainstream were not understood and not supported.

The creators of new, revolutionary ideas in algebra, the Norwegian Niels Henrik Abel (1802–1829) and the Frenchman Evariste Galois (1811–1832), died unknown and misunderstood.[14] The founders of non-Euclidean geometry, the Russian Nikolai Ivanovich Lobachevskii (1792–1856) and the Hungarian Janos Bolyai (1802–1860) also did not live long enough to hear a word of recognition.[15] Especially sad was the life of Janos Bolyai — spent mostly in deep depression.

---

[14]The same can be said about the physicist Carnot. However, his work was pulled from oblivion sooner.

[15]Lobachevskii and Bolyai independently came to the main ideas of non-Euclidean geometry in 1825. Lobachevskii gave a public report on his results in February of 1826. It was not published, as the referees either avoided reviewing or gave negative reviews. Bolyai first published his results in

The stories of Abel, Galois, Lobachevskii, and Bolyai are widely known. There are other sad but less known stories. In 1844, the German mathematician, physicist and Sanscritist, Hermann Günther Grassmann (1809–1877), in his now famous work *Lineale Ausdehnungslehre*, introduced into mathematics the concept of a multidimensional space and developed its geometry. The value of this work was not appreciated for many years. The first admirer of that Grassmann paper among physicists, J.W. Gibbs, wrote in his very interesting paper *Multiply Algebra* (1887):

> This remarkable work remained unnoticed for more than twenty years, a fact which was doubtless due in part to the very abstract and philosophical manner in which the subject was presented.... It is a striking fact in the history of the subject that the short period of less than two years was marked by the appearance of well-developed and valuable systems of multiply algebra by British, German, and French authors, working apparently entirely independently of one another.[16] No system of multiply algebra had appeared before, so far as I know... the appearance of a single one of these systems would have been sufficient to mark an epoch, perhaps the most important epoch in the history of the subject.

Note that Grassmann's work was most general, and in historical perspective, it proved to be viable and important. Today Grassmann's name is often met in different combinations (*Grassmann manifolds, Grassmanians, Grassmann numbers*, etc.) in works on theoretical and mathematical physics. The mathematical theories, the foundations of which were created by Abel, Galois, Lobachevskii, Bolyai, and Grassmann, are extensively used in modern mathematical investigations related to solitons.

Let me mention one more story. Most works of the great Czech mathematician Bernhard Bolzano (1781–1848) on foundations of mathematical analysis, in which he was far ahead of his time, were not published until the 20th century. In the book *Science on Science*, published in 1837, he formulated some basic notions of mathematical logic (this science, fully developed in our century, provides the mathematical foundation for computers). This book, as well as *Mathematical Analysis of Logic* (1847) by the English mathematician George Boole (1815–1864), was not seriously read or understood by their contemporaries.

We will see later that computers played a very important role in the modern history of the soliton. It happened that the first attempts to devise a computer were made at the time of the first observations of the soliton. In August of 1834, when Russell first met his solitary wave, an English mathematician, economist and inventor, Charles Babbage (1792–1871), had formulated the main principles of his "analytical engine." Babbage was far ahead of his time, and his ideas were realized

---

1832 in an Appendix to the first volume of a textbook in mathematics written by his father, Farkas Bolyai who earlier had communicated the work of his son to Gauss, an old friend of his.

[16]Here Gibbs means Hamilton's paper on the so-called quaternionic algebra (1844) and the paper published in 1845 by the French specialist in mechanics and hydrodynamics Barré de Saint-Venant (1797–1886), the subject of which was even closer to the Grassmann multidimensional algebra and geometry.

more than a hundred years later when these principles formed a theoretical foundation of modern computers. We cannot blame Babbage's contemporaries. Many educated men of that time understood the usefulness of "analytical engines" capable of making long and tedious calculations, but the science and technology of that time did not yet allow for a realization of Babbage's bold project on a full scale. The Prime Minister of the United Kingdom, Sir Robert Peel, a graduate of Oxford University with a first in mathematics, was probably sympathetic to Babbage's project, but unfortunately, his conclusion was that financing the construction of a universal computing machine should not be among priorities for Her Majesty's Government. The matter was not, however, closed by this decision and there was a slow but steady development of Babbage's ideas; there were even attempts to realize at least part of them.[17]

Science and Society

> A Frenchman who arrives in London will find philosophy, like everything else, very much changed there.
>
> François Marie Arouet Voltaire (1694–1778)

> What is now the custom in London is still premature in Moscow.
>
> Aleksandr Sergeevich Pushkin (1799–1837)

The achievements and failures of science depend on the economical, technological and educational level of society, and it was not accidental that the main developments of natural sciences in the first half of the 19th century were occurring in Britain, France, and Germany. It is interesting to note that the style of scientific work, the social status of the individuals making science and, more generally, the attitude of society to science were markedly different in those three countries.

In France, scientific activity was strictly organized by the Academy of Sciences. In fact, any work not supported by the Academy, or at least by some respected academician, had little chance of becoming known; it would be difficult to publish such work. In contrast, the work supported by the Academy had very bright prospects. This state of affairs often generated protests from young scientists. For instance, in the paper in memory of Abel, a French friend of his wrote:

> Even in the case of Abel and Jacobi, the benevolence of the Academy did not mean recognizing achievements of these young scientists. It might rather be interpreted as an encouragement of investigations in a certain restricted domain of science outside of which, according to

---

[17]The history of computers is nicely presented in Goldstine *The Computer from Pascal to Von Neumann*, Princeton University Press, 1972.

the Academy's opinion, there can be no progress of science.... We will advise quite differently — young scientists, please listen only to your own "inner voice," don't listen to other advisors. Read the creations of geniuses and meditate on them but never allow yourself to be a disciple without your own opinion.... Freedom of views and objectivity of judgement must be your banner.

Germany at that time consisted of many small states, and scientific life was concentrated around the universities. Most of these universities supported scientific research and scientific journals that were becoming important tools of intercourse among scientists and growing scientific schools. The first scientific schools were born in Germany, and modern scientific journals are in general very similar to these first German journals.

In England the situation was different. There was no academy and no scientific schools. The majority of English scientists worked in solitude. These solitary scientists were rather successful in finding new methods, but their work was often not known to other scientists, either on the continent or in England. In addition, some English gentlemen did not care to send reports of their work to journals; at best, they occasionally reported selected results at meetings of the Royal Society. An extreme example is Sir Henry Cavendish (1731–1810). This rich and eccentric nobleman's hobby was science. He worked in complete solitude and published only two of his spectacular discoveries. His other works were found and published by Maxwell, who devoted a good part of his time to this endeavor. Meanwhile, many of Cavendish's discoveries were rediscovered by other scientists. Sir John Rayleigh was more of an amateur, and performed most of his experiments in his home, but unlike Cavendish, Rayleigh published more than 400 scientific papers and the beautiful book mentioned before. Maxwell was a professional scientist, but he also spent several years working alone in his family home.

We should not misconstrue the solitude of English scientists in too literal terms. Most scientists need to talk regularly to their colleagues. In England a scientist could meet "everybody" at meetings of the Royal Society. Most of the English scientists were personally acquainted with each other and many of them had friendly relations. For instance, we know that Charles Darwin (1809–1882) frequently visited Charles Babbage, who was a friend from their student years of John Herschel (1792–1871), who, in turn, had a friendly relationship with John Russell, etc. The Royal Society was not only a social club for scientists, but it also controlled a very reasonable amount of money for scientific research.

In summary, science developed rather successfully though in somewhat different styles in the three countries. The development was most systematic and well organized in France, the research work was most extensive in Germany, and the most original ideas were advanced in England.

One might wonder how solitary scientists could compete with well organized scientific communities and even surpass them in creating new ideas, sometimes so unusual and surprising that the most qualified experts had difficulty understanding them. To answer this question, let us recall the two greatest figures in the natural sciences — Faraday and Charles Darwin. They worked in very different domains of science and both were completely independent of mainstream thought. Instead, they relied completely on their own minds, their own ingenuity in finding problems, and their own persistence in looking for solutions. If their observations or experiments required it, they would unhesitatingly introduce a radically new concept, unconcerned with whether it could be easily understood by other scientists.[18] It is also important to note that educated society was not indifferent to the scientific quest. Even Babbage, who had not met with understanding of his ideas by others and was gradually becoming a misanthrope, had friends who evaluated his work highly. Charles Darwin and Michael Faraday, for example, held a very high opinion of Babbage's work and ideas. A distinguished mathematician, Lord Byron's daughter Ada Augusta Lovelace, was Babbage's close collaborator and the creator of the first mathematical programs for his computer. He was regarded by his contemporaries as a great but somewhat eccentric man ahead of his time.

By mid-century the social role of science and the importance of science for economic development was clear to many educated people in Britain, and this sometimes allowed scientists to find money for research in spite of the absence of regular government financing. In fact, the Royal Society and the major British universities had more money for research than any leading scientific institutions on the continent. Thus we see that the spiritual and material conditions for the development of science in Britain were not less favorable than in France or Germany. In other countries of Europe, the conditions for systematic development of science in the first half of the 19th century were not so good as in Britain, France, or Germany. Nevertheless, there were brilliant physicists and mathematicians in Italy, Switzerland, Austria and the Russian Empires as well as in other countries.

Scientists in the United States had not yet made significant contributions to physics or mathematics during the first half of the 19th century. Pure science was not popular at American universities, and society was more practically oriented, pragmatic, than, say, in England. There were exceptions to this rule, a well-known one being the work of Joseph Henry, but even Henry was most successful in what we now call "applied" science. Nevertheless, the United States later gave us one of the greatest physicists of that century, Josiah Willard Gibbs (1839–1903).

Note that scientists of the first half of the 19th century were very good in applying new scientific discoveries to technology. Consider, for example, the telegraph, whose foundations had been laid by Ampère and Faraday in their work on electromagnetism. In 1829, Ampère first clearly formulated the idea of telegraph communication. One of the first working models of the electromagnetic telegraph was demonstrated in St. Petersburg in 1832. It has been constructed by the Russian Orientalist and inventor Pavel L'vovich Schiller (1786–1837). In 1836 Schiller constructed a line around the Admiralty building in St. Petersburg, connecting

---

[18]What a contrast to Gauss, who had not published his own results on non-Euclidean geometry and had not written a word in support of Bolyai or Lobachevskii, because he was sure that these ideas would not be accepted by his colleagues!

two of its offices. In 1837 the Russian Government allotted money to construct
a telegraph line between Kronstadt and Peterhoff. This project was not realized
because Schilling died suddenly.

In 1833, Gauss and Weber constructed the first electromagnetic telegraph in
Germany. In 1837, British scientists Charles Wheatstone (1802–1875) and William
Cook (1806–1879) made an even better version of the telegraph. In the same year,
the American painter and inventor Samuel Morse (1791–1872) constructed his
first telegraph apparatus. All of these devices were still unsuitable for commercial
use. The first completely satisfactory commercial version was designed by Morse
in 1840, and in 1844 the line connecting Baltimore with Washington went into
operation.

Chapter 2
The Great Solitary Wave
of John Scott Russell

In the field of observation, chance only favors those minds which have been prepared.

Louis Pasteur (1822–1895)

In view of the general interest in waves, observing a new wave phenomenon in the beginning of the 19th century should not be considered as a purely chance event. But the emotional reaction of the discoverer of the soliton to his discovery was exceptional. Devoting many years of his life to studying its properties and thinking of it, it is evident that the fate of the discovery is closely interwoven with the fate of the discoverer, and so it is of interest to learn something about Russell's life and accomplishments.

The discovery of the soliton is not described in books on the history of physics and mathematics, and you will not find Russell's name in textbooks or in biographical reference books. His first biography was published in 1977 by George S. Emmerson, *John Scott Russell – A Great Victorian Engineer and Naval Architect* (John Murray, London). In spite of a "paucity of primary sources and the perplexing erasures of time," Emmerson produced an excellent biography. We will touch upon only the main events of Russell's life having a relationship to the soliton story.

Before the Fateful Encounter

John Scott Russell was born in Scotland, near Glasgow. He was a son of a clergyman and was supposed to follow his father's profession. However, very early,

A.T. Filippov, *The Versatile Soliton*, Modern Birkhäuser Classics,
DOI 10.1007/978-0-8176-4974-6_2, © Springer Science+Business Media, LLC 2010

the boy showed a strong inclination towards the exact (practical) sciences. So his father let him attend the universities of St. Andrews,[19] of Edinburgh, and of Glasgow, in the latter of which he received his baccalaureate at the age of sixteen.

Such an early start was not unusual at that time. For example, Thomas Young, at the age of 14, had a perfect knowledge of ten foreign languages; at the age of 20, he published his first scientific paper; at 22, he received a doctoral degree in medicine. William Hamilton, at the age of 13, knew thirteen foreign languages; at 16, he read Laplace's *Celestial Mechanics* and even found a serious mistake in it. In Scotland, it was a rule to enter the university at around 16–17 years of age, but exceptional children might start even earlier. For example, William Thomson attended his father's lectures in mathematics at the University of Glasgow at the age of eight! He officially became a student at the age of 10. At the age of 15, he published his first scientific work and, at the age of 21, he graduated from Cambridge University. At 22, he became Professor of Physics at the University of Glasgow. Maxwell entered the University of Edinburgh when he was 16. However, a year before that he had published in the *Transactions of the Royal Society of Edinburgh* his first work — about ovals. At the age of 19, he graduated from the University of Edinburgh and three years later, from Cambridge University. This was also a common practice in Scotland: the "main" universities of the British Isles (at Cambridge and Oxford) were much more prestigious. At the age of 25, Maxwell became a Professor at Aberdeen University – the "fourth" university of Scotland.

Although Russell was clearly a rather gifted young man, his career in science and engineering was not shaping up so smoothly. After the university he spent a couple of years working in factories. Then we find him giving lectures in science at the University of Edinburgh. Russell demonstrated good teaching abilities, and in 1832–1833 he was engaged to read lectures in natural philosophy, when the Chair became vacant after the death of John Leslie (1766–1832).[20] It would be natural for Russell to apply for permanent appointment but apparently he did not, and the Chair went to James David Forbes (1809–1868).

This was not a bad choice. Forbes later became known for his contribution to heat transfer theory and to glaciology. In 1834 he proved by experiments that thermal (heat) radiation may be polarized, like light. He also investigated interference phenomena with thermal radiation and thus established its wave nature. Like Leslie, he constructed several instruments, e.g., the so-called "Forbes's Bar," which was used for measuring thermal conductivity of metals. He was really a good successor of Leslie. In addition, he was probably the first who recognized the talent of young Maxwell. Forbes presented Maxwell's first scientific work to the Royal Society of Edinburgh and later recommended him for a professorship in physics at the University of Aberdeen.

---

[19] This is the oldest one in Scotland, founded in 1411.

[20] Leslie was a well known scientist, a fellow of the Royal Society of London. Most successful were his experiments on heat radiation and absorption for which he invented several original instruments, e.g., the so-called "Leslie Cube."

An academic career was probably not Russell's destiny, although he tried to become a university professor once more. In 1838, the University of Edinburgh announced a vacancy in mathematics. In spite of a very good reference from John Hamilton (he characterized Russell as "a person of active and inventive genius"), he was not chosen. We don't know why. We only know that such decisions usually depend on many different factors, and so talent, activity and ingenuity may happen to be not the most important. In 1857, Maxwell strongly criticized the criteria for selecting scientists in Scottish universities (he blamed the selection system for giving priority to local political consideration rather than to the scientific qualities of the candidates).

We don't know whether this has a direct relation on Russell's failure. However, we know that, in 1860, Maxwell himself applied for the Chair vacated by Forbes in Edinburgh, but it was given to Peter Guthrie Tait (1831–1902). Some of the members of the selection committee strongly criticized Maxwell for being a too original and excessively learned scientist. Tait was probably chosen for being more conventional and closer to the committee concept of a professor in natural philosophy. He later became well-known for his work in physics and mathematics, but his contribution to science is incomparable to that of Maxwell.

Russell was more successful in his career as an engineer, which eventually led him to the soliton discovery. Around 1833, he invented the trolley. This gave rise to the founding of the "Scottish Steam Carriage Company." Although the lifespan of the company was rather short, Russell acquired a reputation as a gifted engineer-inventor. One day, the "Union Canal Society of Edinburgh" asked him about the possibility of applying steam power navigation to the Edinburgh and Glasgow Canal (instead of good old horse power navigation). He replied that the answer to this question required experiments that he was willing to do if some portion of the canal were placed at his disposal. So he was engaged in 1834 in an unpaid study of the design of canal barges, in the course of which his encounter with the solitary wave occurred. He published the first report of this event in 1838. A more detailed account of it and of successive experiments was published in his *Report on Waves* (1844).

John Russell Meets The Solitary Wave

> That impression was imprinted
> Very deep into his heart.
>
> Aleksandr Pushkin

His fateful encounter with the soliton probably occurred near the "Hermiston Experimental Station" on the Union Canal, six miles from the center of Edinburgh. He described it as follows.

I was observing the motion of a boat which was rapidly drawn along a narrow channel by a pair of horses, when the boat suddenly stopped — not so the mass of water in the channel which it had put in motion; it accumulated round the prow of the vessel in a state of violent agitation, then suddenly leaving it behind, rolled forward with great velocity, assuming the form of a large solitary elevation, a rounded, smooth and well-defined heap of water, which continued its course along the channel apparently without change of form or diminution of speed. I followed it on horseback, and overtook it, still rolling on at a rate of some eight or nine miles an hour, preserving its original figure some thirty feet long and a foot to a foot and a half in height. Its height gradually diminished, and after a chase of one or two miles I lost it in the windings of the channel. Such, in the month of August 1834, was my first chance interview with that singular and beautiful phenomenon which I have called the Wave of Translation, a name which it now generally bears: which I have since found to be an important element in almost every case of fluid resistance, and ascertained to be of the type of that great moving elevation of the sea, which, with the regularity of a planet, ascends our rivers and rolls along our shores.

To study minutely this phenomenon with a view to determining accurately its nature and laws, I have adopted other more convenient modes of producing it than that which I just described, and have employed various methods of observation. A description of these will probably assist me in conveying a just conception of the nature of this wave.

Russell performed numerous experiments in natural environments — on canals, rivers and lakes — as well as in his "laboratory," which was a specially designed small reservoir in his garden. In these studies he found the following main properties of the solitary wave.

1. *An isolated solitary wave has a constant velocity and does not change its shape.*

2. *The dependence of the velocity v on the canal depth h and the height of the wave, y_0, is given by the relation*

$$v = \sqrt{g(y_0 + h)},$$

where g is the gravity acceleration. This formula is valid for $y_0 < h$.

3. *A high enough solitary wave decays into two or more smaller solitary waves*: "The wave will assume its usual form... and will pass forward with its usual volume and height; it will free itself from the redundant matter w by which it is accompanied, leaving it behind, and this residuary wave, w_2, will follow it, only with a less velocity, so that although the two waves were at first united in the compound wave, they afterwards separate... and are more and more apart the further they travel." See Figure 2.1 that reproduces Russell's original drawing.

4. *There exist only solitary elevations (humps); solitary cavities (depressions) are never met.*

Figure 2.1 Russell's picture of the decay of one solitary wave into two waves
moving with different velocities in the same direction

Russell once made a remark that "the great primary waves of translation cross each other without change of any kind in the same manner as the small oscillations produced on the surface of a pool by a falling stone." Although he never returned to discussing this really striking phenomenon, the remark once more demonstrates his exceptional observational abilities. Researchers were very much puzzled by this phenomenon, which they saw 130 years later in a very different environment.

One hundred years after Russell's death, participants at the international conference "Soliton–82" held at Edinburgh tried to produce the soliton in the same canal and at the same place where it was first observed. They apparently used a similar barge, but using manpower instead of the horsepower. For this reason or another, the attempt was unsuccessful.

To interpret the results of his experiments, Russell introduced a more detailed classification of waves. He distinguished solitary (lone) and gregarious (group) types of waves and four "orders" of waves (like classification in zoology!). To the gregarious type he attributed the usual waves generated by wind, wave groups, now called *wave packets*,[21] and capillary waves. Russell called his favorite solitary wave the "great" or "primary" translation wave (it is of order one). His classification of different kinds of waves shows that he was not only an exceptionally observant scientist and excellent experimenter but a first class theorist. He was aware of the limitations posed by contemporary mathematics and cautiously concluded his discussion by saying that the solitary wave is "a worthy object for the enterprise of a future wave – mathematician." The attitude to solitons of some most prominent theoretical experts was quite different.

That is Impossible!

If you do not expect the unexpected, you will not find it; for it is too hard to
be sought out, and difficult.

Heraclitus [of Ephesus] (ca 550–475 BC)

---

[21]Now we know that wave packets are also solitons different from Russell's soliton, but to find their solitonic nature by simple observation is practically impossible.

Russell's *Report on Waves* was probably unnoticed on the continent. Unfortunately, the best expert in waves, W. Hamilton was at that time completely dedicated to quaternions[22] and had lost interest in waves.[23] In England the *Report* was read by at least two distinguished scientists – Airy and Stokes.

The Royal Astronomer George Biddell Airy (1801–1892) was a gifted personality. At the age of 22, he graduated from Cambridge University. Three years later he attained a professorship. At 27 he became the director of the University Observatory and at 34 he started his 45 year service as the director of the famous Greenwich Observatory. He contributed significantly to science and, especially, to applications of science. He certainly was one of the most influential scientists in Great Britain.

Airy carefully studied Russell's report and severely criticized his reasoning, in his own paper on tides and waves published in 1845. In this paper Airy stressed that Russell's formula for the velocity of the solitary wave contradicted Airy's theory of long waves on shallow water. He also asserted that no long wave in a canal can preserve its shape. In fact, he argued against all four of Russell's observations and concluded: "We are not disposed to recognize this wave as deserving of the epithets "great" or "primary...."

Unfortunately, such an overcriticism was not atypical of Airy. Just one well-known example. In 1845–1846 the dramatic events related to the discovery of the planet Neptune aroused much controversial and hot discussion among scientists as well as in wider circles of European educated society. Airy's role in these events was rather dubious and he was strongly criticized in England. We recall only the highlights of the story. Airy was one of the first to notice irregularities in the movements of the planet Uranus. However, when John Adams, a young mathematician, gave him the results of his own computations which explained these irregularities as the influence of an unknown planet, Airy made no moves to check the prediction. Only when it became known to him that the French astronomer Le Verrier had reached the same conclusion did Airy give orders to look for this planet. But it was too late. German astronomers, who were not as skeptical as Airy, had found the planet Neptune. This story caused some damage to Airy's scientific reputation. However, Airy was a very consistent conservator. It was also he who had the dubious honor of killing Babbage's analytical engine project. In reply to Prime Minister Peel's request, he characterized the project as absolutely useless.

George Gabriel Stokes (1819–1903) started his work in hydrodynamics when he was a student at Cambridge University. Then he wrote his first scientific paper, which was devoted to planar flows of liquids. During his long scientific career, he studied flows of viscous liquids (the Navier-Stokes equations), many problems of elasticity theory, etc. He is rightly regarded as one of the fathers of modern

---

[22]In 1844, he invented this peculiar system of "numbers" with three imaginary units, which has a close relationship to vectors. Maxwell further developed the theory of quaternions and applied it to his electromagnetic theory.

[23]He had not even published in detail his very important discovery of the *group velocity* of waves which was later attributed to Stokes. We return to this in Chapter 5.

hydrodynamics, but his name is also well known in pure mathematics (the Stokes theorem is known to every student in mathematics or physics). In his paper "On the Theory of Oscillating Waves" (1847) he discussed Russell's observations and theoretical considerations more carefully than Airy. However, his conclusion was that the solitary wave cannot exist even in liquids with vanishing viscosity. This was really bad for the solitary wave.

All the Same it Exists!

After this deadly criticism, the solitary wave was forgotten by everybody except Russell. At the age of 57, he wrote:

> This is a most beautiful and extraordinary phenomenon: the first day I saw it was the happiest day of my life. Nobody had ever had the good fortune to see it before or, at all events, to know what it meant. It is now known as the solitary wave of translation. No one before had fancied a solitary wave as a possible thing. When I described it to Sir John Herschel, he said, "It is merely half of a common wave that has been cut off." But it is not so, because the common wave goes partly above and partly below the surface level; and not only that but its shape is different. Instead of being half a wave it is clearly a whole wave, with this difference, that the whole wave is not above and below the surface alternately but always above it. So much for what a heap of water does: it will not stay where it is but travels to a distance.
>
> *The Modern System of Naval Architecture* (Day & Son, London 1865).

Not long before his death, Russell summarized his work and his reflections on solitary waves and related matters in his book *The Wave of Translation in the Oceans of Water, Air and Ether*. The attitude of 19th century scientists to this book and to the solitary wave in general is clear from the following quotation:

> We need not say more than that the existence of the *long wave* or *wave of translation*, as well as many of its important features, were here first recognized, and (to give one very simple idea of the value of the investigation) that it was clearly pointed *why* there is a special rate, depending on the depth of the water, at which a canal boat can be towed at the least expenditure of effort by the horse. The elementary mathematical theory of the long wave is very simple and was soon supplied by commentators on Scott Russell's work; a more complete investigation has been since given by Stokes; and the subject may be considered as certainly devoid of any particular mystery. Russell held an opposite opinion, and it led him to many extraordinary and groundless speculations, some of which have been published in a posthumous volume, *The Wave of Translation* (1885). His observations led him to propose and experiment on a new system of shaping vessels, which is known as the *wave system*. This culminated in the building of the enormous and

unique 'Great Eastern', of which it has been recently remarked by a competent authority that "it is probable that, if a new 'Great Eastern' were now to be built, the system of construction employed by Mr. Scott Russell would be followed exactly."

Though his fame will rest chiefly on the two great steps we have just mentioned, Scott Russell's activity and ingenuity displayed themselves in many other fields: steam trolleys for roads, improvements in boilers and in marine engines, the immense iron dome of the Vienna exhibition, cellular double bottoms for the iron ships, etc. Along with Mr. Stafford Northcote (now Lord Iddesleigh), he was joint secretary of the Great Exhibition of 1851; and he was one of the chief founders of the Institution of Naval Architects. . . . He died at Ventnor, June 8, 1882.

Encyclopedia Britannica 9th Ed. (1886)

We see that the solitary wave is not recognized here as a particular phenomenon, and that general ideas on solitary waves are called *extraordinary and groundless speculations*. A hundred years later Russell's achievements look different, as one may see from the following excerpt from a textbook on soliton theory.

Scott Russell was very much a man of his time — bold, hardworking and ready to have a go at anything. . . . It was his yard in London that built The Great Eastern, the famous ship designed by Brunel, and the tribulations, accidents and failures associated with this venture have unfortunately, and possibly unfairly, tainted Scott Russell's name. It seems from present day evidence that, in a number of incidents, Scott Russell was used as a convenient scapegoat, perhaps because of his humble background and lack of private fortune. It therefore seems fitting that he should get a large measure of posthumous credit for his original researches into solitary waves. His interest in scientific and engineering problems extended to many other topics, as shown by his extensive list of scientific publications. In particular he made an independent discovery of the Doppler effect. In later life the significance of the solitary wave became almost an obsession, and in his posthumous work *The Wave of Translation* (1885) he applied a theory of solitary waves in air and ether to predict the depth of the atmosphere (correctly!) and the size of the universe (incorrectly). He would have been delighted, but not surprised, to find his predictions of the wide ranging importance of the solitary wave verified in many scientific fields.

R. K. Dodd, J. C. Eilbeck and J. D. Gibbon,
Solitons and nonlinear wave equations.
(Acad. Press, 1982, London)

You see that Russell's life was not dull! I would rather say that it was adventurous, and that the soliton was not the only adventure. Great Eastern was undoubtedly a great adventure. The ship was enormous in size — its iron hull was 207 meters long and 25 meters wide. The two steam engines developed up to 8000 horsepower. In addition to the screw propellers it was supplied with two paddle wheels and six masts. Designing and constructing such a vessel, which had begun in 1853, encountered many problems. I do not know the details and only quote R. K. Bullough, who wrote about Russell's misfortunes related to this great project:

> He entered the first of his periods of bankruptcy before the ship could be launched, became embroiled in the controversy surrounding the failure of a steam valve during the trials of that vessel in the English channel in 1859 and the death of five seamen... in 1867 he was forced to resign from the Institution of Civil Engineers of which he was a founding member following a charge of unprofessional conduct associated with a second financial failure whilst acting to purchase guns for the North in the American Civil War.

There is one more connection of the Great Eastern to our soliton story. In 1865–1866 it was used in laying the telegraph cable on the bottom of the Atlantic Ocean from England to the United States. The dramatic events of this venture were described by the German writer Stephan Zweig in the 20th century.

Rehabilitation of the Solitary Wave

> I'm hymning as before...
>
> Aleksandr Pushkin

Fortunately, Russell lived long enough to see his dear solitary wave acquitted. The younger generation of scientists was less conservative and more objective. In 1871, the French scientist Joseph Valentine de Boussinesq (1842–1929) started anew with the problem by formulating a different approach to the shallow water wave concept. He showed that Russell's solitary wave may exist and approximately calculated its shape and velocity. This conclusion was later confirmed by Rayleigh (1876) and Saint-Venant (1885).[24] These three scientists established a firm mathematical basis for the solitary wave. Russell's observations were also supported by further experiments, his formula for the soliton velocity was exactly confirmed, and the sources of the mistakes made by Airy and Stokes were found.[25] Thus, by the beginning of the 1890s the solitary wave seemed to be firmly established. Nevertheless, the prestige of Airy and Stokes was so high that

---

[24] Rayleigh was one of Stokes' students while Saint-Venant was older than Airy and Stokes. In general, your physical age does not completely determine to which generation you belong.

[25] This was done in 1891 by J. McCowan in the paper "On the Solitary Wave" published in *London, Edinburgh, and Dublin Philosophical Magazine and Journal of Science* that was usually simply called *Philosophical Magazine*. In this paper McCowan also mentioned further experiments on the solitary waves performed in France.

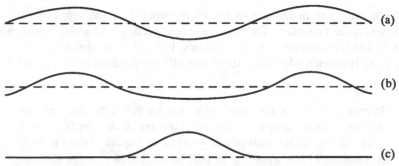

Figure 2.2 Periodic waves (a) linear; (b) nonlinear; (c) a solitary wave

even these further developments appeared to be insufficient to finally establish the truth. Even good treatises on hydrodynamics that appeared after the Rayleigh, St. Venant, and Boussinesq papers still asserted that Airy was correct in negating the solitary wave!

The final resolution of this controversy was given by the Dutch scientist Diderik Johannes Korteweg (1848–1941) and his student Gustav de Vries. They reexamined and summarized all previous considerations, confirmed that Airy's and Stokes' opinions were incorrect, and introduced a new equation for describing solitary waves that we now call the *Korteweg–de Vries (KdV)* equation. These results were published in the paper "On the change of Form of Long waves Advancing in a Rectangular Canal, and on a New Type of Long Stationary Waves," published in 1895 in the *Philosophical Magazine*. The main goal of the paper was to fix once and for all the status of the solitary wave, which was still misunderstood as we have seen above.

For us, more important is another result of this paper. Starting from Rayleigh's and Boussinesq's methods, Korteweg and de Vries found a simpler equation for shallow water waves (the KdV equation), derived their periodic solutions (the KdV waves) and showed how long waves could generate solitons. The KdV waves, like Gerstner's, have non-sinusoidal shape; they are approximately sinusoidal only in the limit of small amplitude, see above Fig.2.2a. When the wavelength is very large, the wave has the shape of two humps (Fig.2.2b), and the distance between them grows with the wavelength. In the limit of infinite wavelength, we have just one hump, which is Russell's solitary wave.

From the mathematical point of view, KdV waves are somewhat more complicated than solitons or Gerstner waves. In particular, they are impossible to represent by elementary functions. The shape of KdV waves can be described only by using *elliptic functions* which are not studied in high school. Elliptic functions were discovered by Abel and later studied by many distinguished mathematicians, especially by the German scientists Karl Gustav Jacob Jacobi (1804–1851) and Karl Theodor Wilhelm Weierstrass (1815–1897).

The KdV equation played a crucial role in our times in resurrecting the soliton. Its importance for physics lies in its ability to describe not only shallow water waves but also many other nonlinear waves. In mathematics, it was a starting point for developing a deep and beautiful mathematical theory. In fact, for pure mathematicians the history of solitons begins with the KdV equation. However, let us not forget that mathematicians did not fully realize the importance of the shallow water equation and neglected the KdV equation until the physicists returned it to life after 70 years of oblivion. An interesting point is that Korteweg and de Vries themselves had no idea of the brilliant future awaiting their equation. Although they called it "very important," they did not go into its detailed discussion and, moreover, they never published anything else on it!

As far as we know, Korteweg during his long and successful career in science[26] did not return to studying the equation. It is not even mentioned in his posthumous biography published in 1945. We do not know much about de Vries' interests and work. It is known that he was a high school teacher of mathematics (like Grassmann) and a member of the Dutch Mathematical Society. After his PhD thesis, which he made under Korteweg's guidance (their famous paper was based on the thesis), he published two papers about cyclones, in 1896 and 1897. I do not know anything about his life and work after that. It appears likely that he never published anything else related to the KdV equation.

Other scientists also showed little interest in the KdV equation, although, from time to time, new generations of specialists in hydrodynamics discussed the solitary wave and its description by the KdV. Thus there were some discussions around 1925, and a more steady interest had set in after the end of the Second World War.[27] In 1946, the Russian mathematician Mikhail Alekseevich Lavrent'ev (1900–1980) gave a first (very complicated) mathematically correct proof of the existence of the solitary wave. In 1954, the American mathematicians Kurt Friedrichs and D. Hyers found a simpler proof. Around that time very precise experiments with solitary waves were performed. A new interesting feature of these experiments was the use of cinema photography. Unfortunately, only a narrow circle of experts in hydrodynamics knew of these developments in the theory of solitary waves.

The Solitary Wave in Solitude

> I know how helpless an individual is against the spirit of his time.
>
> Ludwig Boltzmann

The isolation of the solitary wave from the mainstream of science is not difficult to explain. The main reason is that waves on water are significantly different from

---

[26]He was a well-known scientist and held the Chair of Mathematics at the University of Amsterdam for 40 years.

[27]The state-of-the-art of the KdV equation before the War was summarized in the sixth edition of the popular (and very thick!) treatise by H. Lamb: *Hydrodynamics*, Cambridge University Press, 1932.

other waves. Waves of light, sound waves, and radio waves may be regarded as *linear*, while waves on water are generally *nonlinear*. The meaning of this distinction is the following. Linear waves can be added — recall the Huygens principle, phenomena of interference and diffraction. The exact meaning of this statement is that from two arbitrary waves we can make the third one by simply superimposing them. In other words, we can algebraically add two waves and get a third one. Any wave is approximately linear if its amplitude is small enough.[28] Linear waves are mathematically described by *linear equations* and the algebraic sum of any two solutions of a linear equation is also a solution of the same equation.

On the contrary, the algebraic sum of two nonlinear waves does not satisfy the same equation. So, generally, the superposition of two KdV waves is not a KdV wave; the same is true for Gerstner waves. Such waves satisfy nonlinear equations; for these waves the Huygens principle is not true, and we cannot speak of any simple analogues of interference and diffraction phenomena. This makes the mathematical theory of nonlinear waves much more difficult, but this also makes nonlinear phenomena much more interesting.

Until the middle of the 20th century, physicists mostly studied linear waves, while nonlinear phenomena had been studied for some time by specialists in hydrodynamics. It was also known that sound waves from explosions (gun fire) are very different from normal sound waves. Russell was probably one of the first who realized this. In his book, he remarked that the speed of sound from the fired gun significantly exceeds the normal speed of sound. Later these waves, called *shock waves*, were studied in more detail. But, in the 19th century, the "spirit of the time" was to consider linearity as a fundamental principle and to treat any nonlinearity as an annoying exception. So it happened that nobody tried to look for analogues to solitary waves in other physical phenomena, and the beautiful discovery of Russell did not become popular even after its verification. The systematic study of nonlinear waves in optics, radiophysics, and acoustics began closer to the middle of the 20th century. After uncovering many interesting nonlinear phenomena and creating mathematical methods for their treatment, the attitude of physicists to nonlinear waves gradually changed and the "spirit of the time" became more favorable to resurrecting the solitary wave.

Wave or Particle?

Often a good new name may become a creator.

Henri Poincaré (1854–1912)

---

[28]Quantum probability waves are exactly linear — this is a deep physics principle. Other waves lose their linearity property for large enough amplitudes. Thus, a sound wave, produced by an explosion, is nonlinear if you are close to its source but, at large distances, it is linear. Light waves and other electromagnetic waves are normally linear, but the laser beam is formed by nonlinear waves.

One more circumstance played a crucial role in soliton history. Neither Russell nor other scientists who studied the solitary wave more than a century after him noticed its striking resemblance to a particle. As was mentioned above, Russell was aware of the particle-like property of two colliding solitary waves — after collision both solitary waves preserve their shapes and velocities. Although this fact immediately suggests a particle interpretation of the solitary wave, one may have reason at this point to hesitate since these "particles" are rather peculiar. Their peculiarity was also known to Russell; he had mentioned that *the higher wave is always moving faster!* This contradicts the standard concepts of particle dynamics.

Even more subtle is another property of the collision, which Russell had not noticed. When the higher wave overtakes the lower one, it appears that the first simply goes through the second and moves on ahead. Russell obviously thought that this was true. In reality this is not the case. If Russell had had a video camera he could have seen that the process of the collision is somewhat different. When such waves touch each other, the larger one slows down and diminishes while the smaller one accelerates and increases. When the smaller wave becomes as large as the original one, the waves detach and the "ex-smaller," one, which now is higher and faster, goes forward while the "ex-bigger" one is now moving behind at lower speed. As one can see in Figure 2.3, the observable result of the collision is that *the larger wave appears to have "shifted" ahead of the position that it would have occupied in uniform motion (without the collision), while the smaller one appears correspondingly to have "shifted" backwards* (the positions occupied by the free, non-collided waves are shown in the figure by the broken lines).

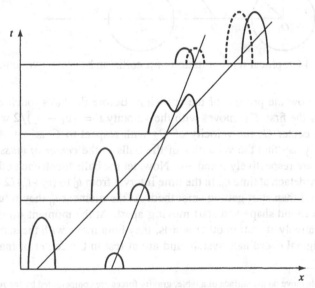

Figure 2.3 Graphical representation of a collision between two solitary waves

Thus, the waves do not penetrate each other, but rather collide like tennis balls. While this is no more than an analogy, it helps us understand the "shift" described above. So, let us consider a collision of two tennis balls uniformly moving along the x-axis. Suppose that the balls are identical, that they do not rotate, and consider their relative motion, see Figure 2.4. Let the velocities of the balls be v_1 and v_2. Then the midpoint O between the centers of the balls moves with the constant velocity $V = (v_1 + v_2)/2$. The point O is obviously the gravity center of the balls (the more precise term is the *center of mass*). As we know from elementary physics, if no external forces act on the balls[29] the mass center moves with constant velocity V along the x-axis.

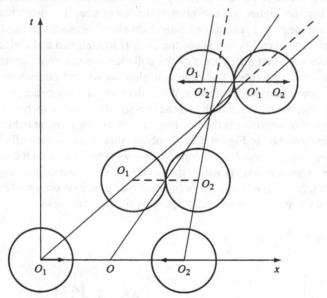

Figure 2.4 Graphical representation of a collision between two tennis balls

Consider now the process of the collision. Before the balls touch each other, the center of the first, O_1, moves with the velocity $v = (v_1 - v_2)/2$ with respect to the mass center O; the velocity of O_2 with respect to O is $-v$. A physicist would simply say that the velocities of the balls in the *center of mass system* of coordinates are respectively v and $-v$. Now, let the balls touch each other at time t_0 and finally detach at time t_0'. In the time interval from t_0 to $(t_0+t_0')/2$ they suffer squeezing and then start getting smoothed. At the moment t_0' they return to their undeformed round shape and start moving apart. At the moment $(t_0 + t_0')/2$ the balls are relatively at rest; in other words, they both move with the same velocity V in the original coordinate system and are at rest in the center of mass system.

---

[29] As the balls move on the surface of a table, gravity forces are compensated by the reaction of the surface. We also suppose that there is no friction between the balls and the table.

Neglecting the issue of energy losses to heating of the rubber, as well as the energy in residual vibrations of the ball's shells, we may conclude that, for $t > t'_0$, the balls move with the velocities $-v$ and v with respect to the center of mass and with velocities $V - v = v_2$ and $V + v = v_1$ in the original system.

Thus the balls exchanged their velocities, but this is not all. As you can see in Figure 2.4, the center of the second ball, O_2, is, at the end of the collision, slightly ahead of the position O'_1, to which the center of the first ball would come if there were no collision. At the same time, O_1 is slightly behind the point O'_2, defined similarly to O'_1. This sort of shift occurs if the time of the collision is smaller than the "characteristic time" $2R/v$, where R is the radius of the balls.

The reader is advised to check that, for a large enough time of the balls interaction, $t'_0 - t_0$, the point O_2 will be behind of O'_1, while O_1 will be ahead of O'_2. We also recommend checking that

$$O_2 - O'_1 = O'_2 - O_1 = 2R - v(t'_0 - t_0).$$

Using this result you can show that $O_2 - O'_1 = O'_2 - O_1 = 0$ if $t'_0 - t_0 = 2R/v$. If the collision time is less than this value, we have the shifts corresponding to those shown in Figure 2.4.

Using these formulas, we can estimate the dependence of the shift on R and on the relative velocity, if we further assume that the collision time is independent of these parameters. When the shift is positive and not small, the dependence is qualitatively similar to what one would observe for collisions of solitary waves. We will later study other solitary waves having shape independent of the velocity. They are even more similar to extended elastic bodies.

All this is simple and clear. Then, why had no one noticed this striking particle-like property of the solitary wave? One thing is obvious. Even such an extraordinarily observant scientist as Russell could not see this subtle effect without proper equipment. More difficult to explain is why experimenters using cinema photography and other modern equipment failed to observe the shift in 1952! The only reasonable explanation of this strange blindness of scientists is that everybody, starting from Russell himself, thought of the solitary wave in terms of wave theory. Though it was clear that this wave is very unusual, nobody could imagine regarding it as a particle.

To some extent, the term "solitary *wave*" itself is to blame for this. In any case, as soon as Zabusky and Kruskal found in their computer experiments that the solitary wave has much in common with particles, they cut off the word "wave" from "solitary wave" and coined the word *soliton*, by analogy with electron, proton and other particles.[30]

---

[30]This happened in 1965. Incidentally, Zabusky and Kruskal first chose the name "solitron" but this word happened to be the registered name of a certain firm. For this reason they were advised to delete the "r" from the name and thus the solitron became *soliton*. I do not know of the fate of the firm, but we all know that the Soliton is flourishing. It was really a lucky name!

Chapter 3
Relatives of the Soliton

In philosophy, it is right to look for similarity even in things looking remote.

Aristotle (384–322 BC)

While the solitary wave was struggling for recognition, other interesting events related to solitons were happening. One occurred in hydrodynamics, another in physiology. Although the first of the discoveries was theoretical and the second experimental, the discoverer of both was one person — Hermann Helmholtz.

His first fundamental contribution to physiology was measuring the velocity of the *nerve pulse*. During the years after this basic work, a number of scientists studied many details of propagating pulses in nerves and even measured their shape. However, only recently has it become clear that *the nerve pulse is a peculiar solitary wave*, having much in common with the Russell soliton.

In another work, Helmholtz discovered that the vortices (eddies) in ideal water[31] have unusual properties making them similar to particles. More precisely, he showed that the vortices in ideal water are indestructible and that they interact like electric currents! Using modern terminology, he demonstrated that the vortices are *soliton-like objects*.

---

[31] If we neglect internal friction, which defines the viscosity of water, and also neglect its compressibility, we get so called ideal water. Richard Feynman colorfully called water without viscosity "dry water."

A.T. Filippov, *The Versatile Soliton*, Modern Birkhäuser Classics,
DOI 10.1007/978-0-8176-4974-6_3, © Springer Science+Business Media, LLC 2010

Hermann Helmholtz and the Nerve Pulse

> At that time physics was not among lucrative employments, and my fa-
> ther declared that he could support my study of physics... if I would take
> medicine as well. I had nothing against investigating animate nature and
> readily consented to his proposal.
>
> Hermann Helmholtz

So started the journey of one of the great scientists of the 19th century into science. His work had significant influence on the development of physics, mathematics, physiology, psychology, and medicine. In his lifetime, he achieved fame
not only in his native Germany, but also in England, France and in other countries where natural sciences were developing. Helmholtz's scientific work had the
best features of the German, English and French scientific schools, and he had
many pupils from different countries. Some of his students later became distinguished scientists and founded their own scientific schools. Among them were
physicists Heinrich Hertz and Piotr Nikolaevich Lebedev, Russian physiologist
Ivan Mikhailovich Sechenov, and many others. Sechenov left us the following
portrait of his beloved teacher:

> His quiet figure emanated a sort of unworldly peace... My impression of
> him was somewhat akin to what I felt while looking at the *Madonna Sistina*
> in Dresden.

Such a remarkable impression of Helmholtz's personality was undoubtedly the
result of his spiritual work, which started in childhood. This work was not easy
and required a strong will to overcome "internal" (weak health) and "external"
(financial) problems. To appreciate this, let us become acquainted with his not
easy but happy life.

He was born on August 31, 1821 into the family of a schoolteacher. His wife
belonged to an English family that had migrated to Germany. Among his ances-
tors were also Frenchmen. While he certainly was a patriotic German, he was
not nationalistic, which is why perhaps he had so many students from different
countries.

But the students, the fame and all the rest came later. The first steps were very
difficult. Seventy-year old Helmholtz recollected his childhood as follows:

> Until the age of seven, I was a sickly boy.... Early, a certain shortcoming of
> my mind also became apparent: a weakness of my memory to things having
> no logical relationship.... I was worse than others in memorizing vocables,
> irregular grammatical forms, and, especially, locutions.... It was a torment
> for me to learn by heart prose works...

However, the attitude of this unusual boy to mathematics and physics was very
different.

With great diligence and joy I was reading all textbooks of physics in my
father's library.

This passion for physics determined his future. In fact, physics always dominated his interests, and even his famous achievements in the physiology of vision
and hearing have their origin in physics.

To the credit of his teachers, they correctly diagnosed young Hermann's talents
in physics and mathematics. In spite of his difficulties with "vocables" he successfully graduated from high school and entered the Military Medical Institute
in Berlin. Influenced by the well-known physiologist Johann Müller (1801–1858)
he started to work in histology and physiology. In 1842, he completed his thesis
On the Texture of the Nervous System of Invertebrates. At that time, physiologists
identified nerve cells and fibers, but a relationship between them was not clearly
understood. Young Helmholtz was one of the first to understand that cells and
fibers constitute whole entities, one of which is now called the neuron.

In 1843, Helmholtz was appointed military physician in Potsdam near Berlin.
There he wrote his famous work on energy conservation. Although at that time its
significance was not widely recognized, he became known as a first-rate scientist,
and, in 1849, Königsberg University gave him a professorship in physiology. He
worked there until 1855, and there he made his celebrated experiment, in which
he measured the velocity of the nerve pulse.

From then on Helmholtz devoted more and more time to research. His fame
continued to grow. Many universities offered him good positions. In 1855, he
was invited to work at Bonn University and in 1858 at Heidelberg University.
His *Science-Popular Lectures* (*Populäre wissenschaftliche Forträge*), first published in 1865–1870, were disseminated all over the world and served several
generations of science students as a very clear introduction to modern problems
of the natural sciences. In 1871, he had to make a choice between a professorship in physics at Berlin University and the Chair of Experimental Physics at
Cambridge. Helmholtz chose Germany and the Chair at Cambridge was given to
Maxwell. From 1888 he was the President of the first Physical-Technical Institute
in Charlottenburg (near Berlin). He held a powerful position in the world of natural sciences and his influence was notable. Educated society regarded him as the
leader of contemporary science. Accordingly, he had more administrative duties
and more social responsibilities, but never stopped his work in science. He passed
away on September 8, 1894.

Helmholtz contributed significantly to physics, hydromechanics, mathematics,
anatomy, physiology, and even to psychology. This was a rare creative breadth
and depth, even for the 19th century, when breadth and depth were not unusual.
Besides the paper on energy conservation, his works with the most direct relation
to our topic were on vortices, the origin of sea waves, and determination of the
velocity of the nerve pulse. For many years he was interested in electromagnetism.
In a lecture devoted to the physical ideas of Faraday (1881), he put forward the
idea of the elementary electric charge. He called it "the smallest possible" electric

charge, or, the "atom of electricity." Recall that existence of the elementary charge was experimentally proven only in 1897.

Now, let us return to the nerve pulse. When, in 1849, Helmholtz began his experiments with the nerve pulse, not much was known of its nature. Scientists believed that the nerve pulse was a sort of electric signal that propagates with enormous velocity, inaccessible to direct measurement. The idea of the electric nature of nerve signals is very old. As I mentioned in Chapter 1, Luigi Galvani stumbled on it while experimenting with frog's nerves. This story is widely known and I will not repeat it. Not so widely known and interesting is that this correct idea was rejected and forgotten for decades due to the clever and beautiful experiments of Alessandro Volta. His experiments had laid the foundation for the development of the theory of electricity which was briefly described in Chapter 1. It was generally believed that Volta also demonstrated that the electricity in Galvani's experiments was not of "animal" origin, contrary to Galvani's ideas.

Galvani's ideas returned to circulation in the first half of the 19th century. A friend of Helmholtz, Emile Du Bois Reymond (1818–1896) contributed to this resurgence. He described his experiments and carefully discussed their meaning in his extensive treatise *Investigations of Animal Electricity* (1848).[32] Investigations performed by Du Bois Reymond and others clearly demonstrated that animal electricity exists and that the nerve pulse is indeed an electric signal. Accordingly, scientists started to regard nerve fibers as electric or telephone wires.

Du Bois Reymond acquainted his friend with these ideas, and Helmholtz was so interested that he decided to check them in experiments. I will not describe his experiment — by Faraday's standards, its design was rather complex, but conceptually, it was simple and beautiful. The result of the measurement looked stunning — the velocity of the pulse propagation in the frog's nerve was not "enormous," it was only 30 m/s ≈ 100 km/h! To Helmholtz's supervisor it looked so unbelievable that he recommended his favorite pupil not to publish the doubtful result, believing there had to be some nontrivial hidden error.

Nevertheless, the result was published and, in due time, accepted by the scientific community. It left a strong and lasting impression. The Russian physiologist, Ivan Mikhailovich Sechenov (1829–1905) later reflected on the lectures by Du Bois Reymond, given in 1857:

> To me, most memorable were his lectures on the rapidity of the excitation propagation along nerves. Here he was speaking with particular animation and told us the whole story of this discovery: doubts of Müller... his own ideas... and, finally the solution of the problem by his friend Helmholtz.

---

[32]Note that Faraday was also interested in this subject, especially, in the electricity of the nervous system.

Further Adventures with the Nerve Pulse

Helmholtz's experiments had shown that a naive concept of the nerve fiber was wrong. However, nobody could suggest anything better, and the real mechanism behind nerve pulse propagation was discovered 100 years later. We will learn of modern developments in Chapter 7. Here, only highlights of the 19th century history are given. Helmholtz returned, from time to time, to studying the nerve pulse. In 1867–1870 he measured the velocity of the pulse in human nerves (in collaboration with N. Bakst of Russia). He apparently was thinking about the nature of the pulse but failed to find a convincing theory.

In 1868, Julius Bernstein succeeded in determining the shape of the pulse. In remarkable experiments, Bernstein also confirmed Helmholtz's result and measured time dependence of the electric potential in the nerve. This dependence proved to be described by a uniformly moving "bell-shaped" pulse. He also found that the velocity of the pulse is equal to the speed of the signal measured by Helmholtz.[33] Later investigations showed that the "bell" *always moves with the same velocity and has the same shape, independent of the original excitation that produced the pulse.*

These remarkable experimental achievements stimulated attempts to create a theory of nerve signals. In 1879, German physiologist Ludimar Hermann (1838–1914) (a former student of Du Bois Reymond) suggested an idea that proved to be rather close to a modern mathematical treatment of the nerve pulse. He suggested that pulse propagation is analogous to flame propagation along the Bickford fuse. For a given fuse, the shape and the velocity of the "solitary wave" running along the fuse is constant (it is $v = m/M$, where M is the mass of the gunpowder per unit length of the fuse and m is the mass that is being burnt in unit time; normally, for the Bickford fuse $v = 1 cm/sec$).

Hermann later proposed a more realistic model of the nerve fiber. He suggested that the fiber might be regarded as a peculiar telephone cable, in which the propagating waves strongly interact with each other.[34] In the beginning of the 20th century, Hermann realized that the wave propagation in such a cable must be described by a nonlinear equation (the "nonlinear diffusion" equation), but mathematicians at that time were unable to solve corresponding mathematical problems.

About the same time, a new "physical-chemical" theory of the processes inside the nerve fiber was advanced by Bernstein. He developed the so-called "mem-

---

[33]Above, we made no distinction between the "pulse" and the "signal." But only after Bernstein's work did it become clear that the nerve signal is in fact a very simple pulse.

[34]The fundamental distinction between propagation of waves in the standard telephone cable, on the one hand, and the wave of flame or the nerve pulse, on the other hand, is the following: The electromagnetic wave can go through the normal cable without any external energy supply, although it will gradually fade due to electric resistance of the cable. The flame and the nerve pulse require a continuous energy supply for their existence. If, in some part of the fuse, there is no gunpowder, the wave of flame will stop. In addition, the waves of flame, unlike electromagnetic waves, cannot be superimposed. Indeed, two colliding waves of flame will extinguish each other and this is often used in extinguishing forest fires.

brane hypothesis" of the nerve pulse, which will be briefly described in Chapter 7. This hypothesis is very close to the modern theory of the nerve pulse, but Bernstein's contemporaries were not enthusiastic about his work.

In summary, the story of the nerve pulse was very similar to the story of Russell's soliton. Once in every twenty or thirty years there was a significant step forward, but nobody knew about these developments outside of a very narrow circle of experts. The result was that, for almost a century, nobody noticed the relation between the nerve pulse and Russell's solitary wave.

Hermann Helmholtz and Eddies

It was Helmholtz who demonstrated rather remarkable characteristics of the vortex motion... starting from the properties of the vortex rings Sir W. Thomson suggested a new version of atomic theory...

Ludwig Boltzmann

The nerve pulse is not similar to a particle. Vortices (eddies) and, especially, vortex rings behave like interacting particles. Their remarkable and even puzzling characteristics were discovered by Helmholtz in 1858. Eddies in water are familiar to everybody, like waves. They are also as easy as waves to produce — if you move your hand in water fast enough, you may produce many of them. Everyone has seen eddies behind oars or boats; also well known are the rings of tobacco smoke. Most of us have seen (at least, on TV) powerful atmospheric vortices — tornadoes. You probably have a good general idea of the *vortex*.

The French mathematician and philosopher René Descartes (1596–1650) regarded vortex motion as a basic property of matter. In his *Foundations of Philosophy* he gave a picture of the Universe full of vortices that run through it everywhere. In Figure 3.1, taken from his treatise, you may see the centers of big vortices – S, F, etc. (particles denoted by dots are rotating around their axes and in turn rotate around S, F, etc.). By the way, Johann Kepler (1571–1630) also thought of the solar system in terms of a vortex that scattered the planets around the sun.

This "Cartesian" picture of the world was rather speculative, and Newton was quite right in his criticism of it. He argued that space is mostly empty and pieces of matter in space interact via long range forces (the generic example was gravitational interaction). As Newton's views were in better agreement with observations, the Cartesian picture was gradually forgotten. Unfortunately, the Newtonian picture of the world was also not free of speculative assumptions (such as absolute space and time, propagation of interactions with infinite speed) that later hampered the development of physics no less than the Cartesian picture. Paradoxically, the Cartesian view of the world was in some ways closer to modern concepts. According to Descartes, there was no such thing as "empty space:"

It is contrary to reason to say that there is a vacuum or a space in which there is absolutely nothing.

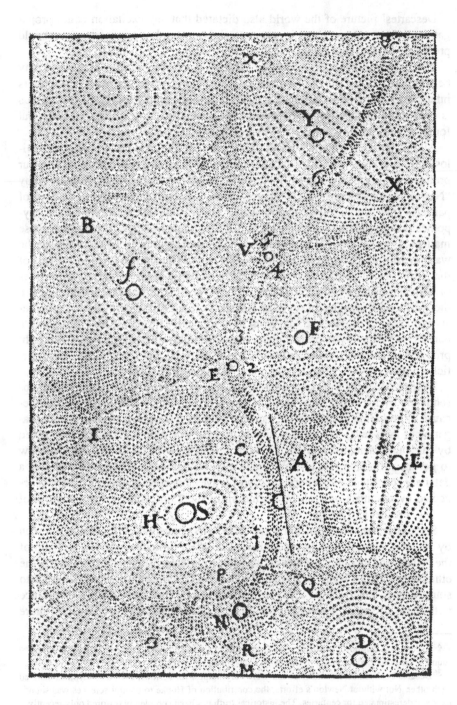

Figure 3.1 A vortex structure of the world according to Descartes

Descartes' picture of the world also dictated that any excitation could propagate only from one vortex to its nearest neighbors, and so any interaction should propagate with a finite speed.

The essential distinction between Cartesian and Newtonian approaches may be summarized as follows: Descartes — no empty space and nearest neighbors interactions. Newton — action at a distance through empty space. Of course, this is over-simplification. In fact, Newton's analysis was more subtle and deep, but for now we need not go into more detailed comparisons.[35]

All this might seem too "philosophical" but, in fact, it is rather a "natural philosophy," meaning that the concepts of space, time and interaction dictate our preferences in approaching more concrete physics problems. Thus, influenced by "Newtonian philosophy," Newton was inclined to believe that light is a flow of particles (the "corpuscular" hypothesis of light as distinct from the "wave" hypothesis), although he was very careful in his published statements about these matters. For example, while discussing light propagation in material media, he was not excluding some periodic (wave?) processes accompanying movements of the particles of light. Moreover, he tried to sketch a theory unifying the virtues of both corpuscular and wave points of view. In fact, he was the first to measure the wavelength of light. In reply to a sharp critique of the corpuscular hypothesis expressed by R. Hooke,[36] Newton tried to clearly formulate his ideas about a particle-wave duality in optical phenomena. Roughly speaking, he regarded light propagating in empty space as consisting of particles, but thought that these particles produce waves when hitting material media (like stones in water).

Newton's view of light phenomena was very complex and over-simplified subsequently by his colleagues. Newton was critical of the wave theory of light because of problems in explaining standard geometrical optics (straight line propagation of light rays). Hooke was unable to solve this problem. It was later solved by the Dutch scientist Christian Huygens (1629–1695), who also showed how to get the refraction law into the wave picture. Although Huygens developed a detailed, consistent and general theory of waves, his wave theory of light was accepted by scientists only at the beginning of the 19th century after the work of Young, Fresnel and Arago.

Soon after the victory of the wave theory, its conceptual difficulties, foreseen by Newton, became widely recognized. The main difficulty was the problem of the "ether," the medium through which light waves must travel. At first glance, the ether seemed to be something like an ideal liquid, with light waves analogous to sound waves. However, the then known properties of light waves were at variance with those of sound waves. For this reason, physicists began to consider more

---

[35] A deep and careful but still popular discussion of these and other concepts in physics may be found in the book *Evolution of Physics* by A. Einstein and L. Infeld.

[36] Debates between Hooke and Newton were very helpful for science, but both scientists disliked each other. Not without Newton's efforts, the contribution of Hooke to natural sciences was significantly underestimated for centuries. The historical truth has been completely restored only recently.

complex movements in liquids and eventually returned to a kind of Cartesian vortex picture.

To my knowledge, this picture was best used by Maxwell in a series of papers "On Physical Lines of Force," published in 1861–1862. These papers might also be called "A theory of molecular vortices and its applications to electric, magnetic and optical phenomena." Another physicist, Ludwig Boltzmann (1844–1906) later wrote a detailed commentary on these works in which he said that this series of papers is one of the most interesting publications in the history of physics.

The "Maxwellian vortex picture" is shown in Figure 3.2, which is taken from his work. In this figure, the arrows A, B represent the electric current, the hexagons represent vortices, and the small circles (small wheels) represent electricity (this closely follows Maxwell's explanations). The axes of the vortices are directed along the lines of the magnetic field, and the angular velocity of the rotating hexagons is proportional to the strength of the magnetic field. When there is an electric current, the wheels make the nearest hexagons rotate, and the rotation propagates to distant hexagons and wheels. So we see that the electric current generates a magnetic field, which spreads in space with a finite velocity. Moreover, Maxwell derived from this model the equations describing the interaction of the electric current with the magnetic field and propagation of the electromagnetic field in space! Having these equations, you may forget about the model, which may be considered as a ladder that you can throw away once you are "at the top." I think that it is this brilliant jump from a rough mechanical model to a most refined and abstract theory that so delighted Boltzmann.

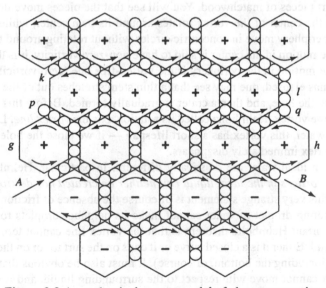

Figure 3.2 A mechanical vortex model of electromagnetism
according to Maxwell

However, Maxwell was not satisfied with this derivation, and later he tried to deduce his celebrated equations using different, more abstract considerations. He invariably stressed that the mechanical picture is nothing more than a model, yet he tried and tried to construct other mechanical and hydrodynamical models of electromagnetism. Most often, he compared electromagnetic processes with vortex motions in the ideal liquid. He was very familiar with Helmholtz's paper on vortices and often quoted it.

This paper was published in the *Journal of Pure and Applied Mathematics*, edited by a German engineer and amateur mathematician August Leopold Crelle (1780–1855). This journal was popular among mathematicians and physicists and usually was called simply the *Crelle Journal*. Crelle especially supported young talented scientists; for instance, he published, in the first issues, Abel's papers. As a result of this wise policy, many papers published in this journal became famous.

Now, let us return to the Helmholtz paper about vortices in the ideal liquid (having zero viscosity and being incompressible). This liquid is mathematically described by the Euler equation. Stokes was the first to divide all motions of this liquid into those without "vorticity" (rotationless, vortex-free, or laminar) and with "vortex motions" (in which there are domains with rotations). He considered in detail only rotationless motions, and so Helmholtz decided to study properties of vortices. The results of his investigation proved to be exciting.

To understand Helmholtz's results, we first have to learn something about vortices and vortex motions. Vortices in water are usually short-lived. So let us first look at the vortex in a bathtub, which is produced by the water flowing out of it. To do this, let us fill up the bathtub and wait until the water becomes quiet. Then gingerly remove the plug and put on the water surface, in the vicinity of the vortex, several short pieces of matchwood. You will see that the pieces move differently. The one in the center of the vortex rotates around its own center, while those in the vortex periphery move in concentric circles without rotating around their own centers. The motion in the center is said to have non-zero vorticity, it is the vortex motion. The motion in the periphery is the vortex-free one (zero vorticity). After some time has elapsed, one may see that a thin stem stretches out of the center of the vortex to the hole, and then a crater is gradually formed. Before this moment, we saw a lone vortex. Helmholtz called its rotation axis the *vortex line*. Like other vortices in water, this vortex has a short lifespan — if we close the hole with the plug, the vortex immediately disappears.

Motions in ideal water are drastically different. In particular, Helmholtz had shown that *vortices in the ideal liquid can neither be created nor destroyed*! The reason for this very strange statement is of course the absence of friction — without it, a rotating droplet of water cannot make its neighbor droplets rotate. The second important Helmholtz theorem is that the vortex line cannot terminate inside the liquid. Either it is a closed curve or it ends on the surface or on the walls of the vessel (including the bottom, of course). It must also be obvious that any isolated vortex cannot move with respect to the surrounding liquid, and it is carried by the stream like a boat.

Even more interesting is *coupling of two vortices*. If two vortices rotate in the same direction (see Figure 3.3), their centers necessarily rotate around a fixed point O situated between them (if the vortices are identical, then $O_1O = O_2O$). When the vortices rotate in opposite directions (one clockwise and the other counter-clockwise), the point O lies outside the segment O_1O_2. Very interesting things happen if these vortices just counter rotate but otherwise are identical (like mirror images). In this case, both vortices move with the same speed like a rigid body (formally, when $v_1 \to v_2$ the point O goes to infinity). To be more precise, all droplets of water inside an oval curve (the boldface line in Figure 3.4) move with the same velocity with respect to the rest of the water. In the rest system of this oval we will see that the rest of the water flows around the oval as if it were a rigid body. This remarkable phenomenon was studied in detail by Kelvin in his paper "On Vortex Motion" (1869). Accordingly, we will call this particle-like pair of vortices *Kelvin's oval*.

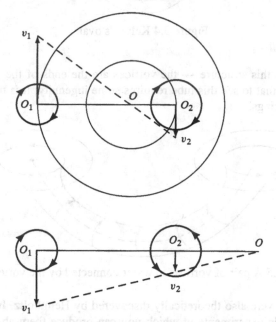

Figure 3.3 Rotation of two vortices with centers O_1 and O_2

Observing a pair of vortices in real water requires considerable patience. You may try to make it by a smooth, short motion of a small scoop on the water surface (don't duck it deep into water!). To observe the vortices it is easier to look at their shadows on the bathtub bottom (after some experiments with the lighting you will succeed in this!). Don't be disappointed if you don't see the oval, but you will surely see the pair. In shallow water this pair disappears very fast, but if the bathtub is full of water you will not only see the pair of vortices but might succeed in observing the structure shown in Figure 3.5. In fact, in deep water the

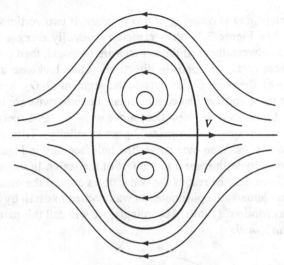

Figure 3.4 Kelvin's oval

pair always has this structure — the vortices are the ends of the *vortex tube*. I must warn you that to see this tube requires some ingenuity; it is much easier to observe vortex rings.

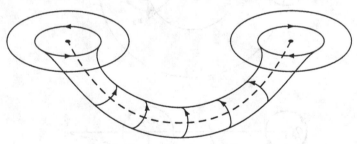

Figure 3.5 A pair of vortices in water connected by the vortex tube

Vortex rings were also theoretically discovered by Helmholtz. In addition, he described simple experiments in which you can produce them abundantly (this is just a scientific approach to producing tobacco smoke rings). Like the pair of vortices, the vortex ring must always be moving; it is not difficult to guess that it moves perpendicular to its plane. One may also convince oneself that the smaller ring moves faster. A beautiful phenomenon may be observed when two identical rings with centers on the same axis OO' are not very far from each other. It turns out that such a configuration (identical rings moving in the same direction with the same velocity) is unstable. The rings are attracted to each other — ring 1 decelerates and grows in diameter, ring 2 accelerates and diminishes, Figure 3.6a. At some moment ring 2, being small enough and fast enough (Figure 3.6b), flies through ring 1, see Figure 3.6c. Then the original configuration repeats itself, with rings 1 and 2 exchanged, and the whole game starts anew. In the ideal liquid this game continues indefinitely.

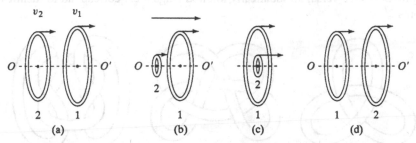

Figure 3.6 Coupling of two vortex rings (a,b,c,d show a sequence of relative positions of the rings 1 and 2)

Helmholtz and Kelvin found these beautiful effects by pure theoretical (mathematical) considerations starting from the Euler equations for the ideal liquid. They did not try to find what happens if the rings move in different directions. Such problems became accessible to a theoretical treatment only recently, with the aid of big computers. The solutions obtained by computers, of which we will talk in more detail later, reaffirmed that vortices have much in common with particles. Let me stress once more that these "vortex particles" have rather unusual properties, and their dynamics are somewhat different from Newtonian dynamics. Thus, instead of Newton's first law we have the following. Any isolated vortex is at rest relative to the surrounding liquid, while Kelvin's oval or Helmholtz's ring always move with constant velocity. Note that two vortices do not interact like normal particles. Two pairs of vortices interact like two extended deformable bodies; the same is true for the vortex rings.

We see that the rings and the pairs of vortices are particle-like objects or, if you wish, soliton-like objects. To some extent, the isolated vortex may also be considered as an extended particle (with some internal "structure" given by the rotation; we might call this object a "spinning" extended deformable particle). We return to a more detailed discussion of this analogy at the end of the book.

Lord Kelvin's "Vortex Atoms"

The particle-like behavior of vortex rings was well known to Kelvin and, using this analogy, he suggested an interesting model of *vortex atoms*. The main idea of the model is the following. Suppose that the universe is filled with an "ether" that is an ideal liquid. Then, if in the act of the creation of the universe there were created vortex rings, they will never disappear. Kelvin suggested regarding these particle-like objects as the most fundamental, primary building blocks of the universe, i.e., atoms. Especially beautiful is the idea of how to visualize different kinds of atoms. Kelvin considered them as different rings having a different number of *knots* (that are impossible to untie). Thus any simple ring corresponds to an atom that is not identical to the atom corresponding to the knotted ring shown in Figure 3.7. Vice-versa, all identically knotted rings must correspond to identical atoms.[37]

Figure 3.7 Kelvin's idea of vortex atoms. Differently knotted
vortex rings represent different atoms

Kelvin could not relate his vortex atoms to any observable particles, and I think that he did not seriously try to find such a relation. Most probably, he considered his theory as a speculative model not directly related to the real physical world. So it is quite natural that his theory was forgotten for decades. In the 20th century, after the notion of the ether was expelled from physics, models of this sort were naturally regarded as completely anachronistic. However, recently some cognate models appeared "on the market" and we will describe some of them later. We now understand that Kelvin attempted to make a first *topological soliton model of elementary particles*. In this sense, his model proved to be quite enduring.

Ending this short discussion of the very interesting idea to which we return later, I would like to quote the Russian physicist Nikolai Alekseevich Umov (1846–1915). He said in 1896 in his speech dedicated to the 300 year anniversary of Descarte's birthday:

It is probable that in the world of thought, like in our material world, no spontaneous germinating exists; there is only an evolution, and the modern

---

[37]Using modern terms, Kelvin classified rings and atoms according to their *topological* properties. *Topology* is a branch of geometry that studies properties of figures that are preserved (invariant) under arbitrary continuous deformations. The number of knots on a closed curve is a topological invariant. Topological considerations are more and more popular in modern physics. We will return to some generalizations of Kelvin's topological ideas at the end of the book.

thought arises (maybe not consciously) on a background of concepts and ideas left to us by our predecessors.

Lord Ross and Vortices in Space

The first encounter with cosmic vortices happened to occur about the same time. In 1848 the Irish amateur astronomer William Parsons (1800–1867, also known as Lord Ross) made the world's biggest telescope-reflector at the time. It was 18 meters long and had a metallic reflector of 182cm in diameter. It was a very impressive apparatus. Even more impressive were his observations of the *spiral structure* in the nebula M51 (Canis Major, or Large Dog). Subsequent observations of Ross and of other astronomers revealed spiral structures in many other nebulae (Figure 3.8).

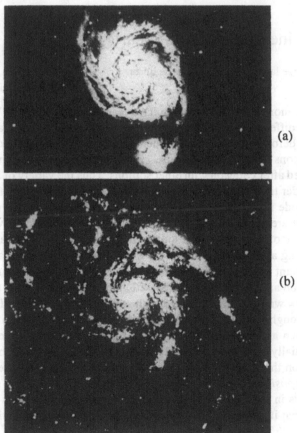

(a)

(b)

Figure 3.8 Photos of typical spiral galaxies

Today we know that nebulae are in fact giant galaxies consisting of myriads of stars as well as interstellar dust and gas. The majority of galaxies have, like our galaxy, a spiral structure and thus may be likened to giant vortices in space. The idea that the nebulae are rotating, so that the vortices in space are not unlike those in water, did not escape Ross' notice. He noted that stars in the spirals are like bodies in rotating vortices, but then remarked that space is so different from water that the dynamical origin of nebular vortices must be different from vortices in water. He further commented that nebular vortices look more like spirals arising with rotations in strongly viscous liquids, but finished his discussion with a somewhat pessimistic conclusion that all such analogies lead to a dead end.

In a historical perspective, this pessimism is quite understandable, but in fact modern science completely justified these analogies. Different kinds of rotations leading to vortex structures are rather typical objects of cosmological science. Maybe the story of this branch of science is as interesting as the soliton story. It is also full of surprises, dead ends, and misdirection. We will not go into details here and will return instead to our soliton journey.

About Linearity and Nonlinearity

> True laws of Nature cannot be linear.
>
> Albert Einstein

Let us once more reflect on what we have learned about solitons. We know now that in very different material media there may exist localized excitations strongly resembling deformable particles. Mathematicians usually prefer to distinguish between "solitons" and "solitary waves." They call *solitons* solitary waves that are not deformed after collision with other solitons. Thus the variety of *solitary waves* is much wider than the variety of these "true" solitons. However, it is hardly natural to include in the variety of solitary waves vortices or vortex rings. For this reason, they are sometimes called "soliton-like excitations." To avoid this awkward term we often use the term soliton in all cases. This is not dangerous when we are talking about general properties of soliton-like objects. In addition, some most important solitons will be discussed below in more detail, and I hope that their common and distinctive features will gradually become clear to the reader.

Up to now we became acquainted with solitons of three types. They were discovered at roughly the same time but nobody suspected any relation between them for more than a century. The deep reason for this was that all these excitations were essentially *nonlinear*, while the general theory of waves in material media was based on the concept of *linearity*. This concept was formulated as a *principle of superposition* of waves and other excitations. In view of its generality and effectiveness in solving diverse physics problems, many scientists were inclined to believe that it is one of the fundamental principles of mathematical physics.[38]

---

[38]Some obviously nonlinear problems of hydrodynamics were, in this sense, annoying exceptions.

Thus a dominant approach to any physical problem was to "linearize" it, at least giving an approximation. This means to find an approximate linear equation satisfying the principle of superposition. In addition, there exist general methods for solving linear equations, while nonlinear equations do not allow so general a treatment. In some very particular cases, special nonlinear solutions were found, such as Gerstner waves, KdV solitons, or hydrodynamic vortices, but these exceptions only stressed the rule — no effective approach existed for solving nonlinear problems.

Let us try to understand by very simple examples why this is so. Consider the simplest algebraic equation, $ax + y = 0$, where a is some fixed number and x, y are two unknowns. Its solutions are pairs (x, y) identically satisfying the equation. In this trivial case, it is obvious that any solution has the form $(x_0, -ax_0)$ where x_0 is an arbitrary number. In geometric terms, the general solution may thus be represented by the points in the (x, y) plane that lie on the straight line OA passing through the points $O = (0, 0)$ and $A = (1, -a)$. Now, linearity means that once we know one solution, say, $A = (1, -a)$, then $x_0 A$ is also a solution. In more general terms, if $A_1 = (x_1, y_1)$ and $A_2 = (x_2, y_2)$ are two arbitrary solutions, then any *linear combination* of them, $A_3 = c_1 A_1 + c_2 A_2$, is also a solution. This means that for any values of c_1, c_2 the point A_3 also lies on the straight line OA.

Mathematicians say that the totality of the solutions of the equation form a *linear variety* or *linear manifold*. As all the solutions are obtained from just one by multiplying it by any number x_0, this manifold is *one-dimensional*.

Quite similarly, we may convince ourselves that all solutions of the equation $ax + by + z = 0$ form a two-dimensional linear manifold. In fact, any solution is a linear combination, $c_1 A_1 + c_2 A_2$, of two independent solutions, $A_1 = (1, 0, -a)$ and $A_2 = (0, 1, -b)$. Geometrically, this is a plane embedded in three-dimensional space (try to prove all these statements and to draw the plane for special values of the parameters a and b).

Returning to physics, let us recall a high school physics problem. A bob of given mass hanging by a spring is known to oscillate around its equilibrium position. Denoting the deviation from the equilibrium position at the moment t by $x(t)$, we may write for any oscillation $x(t) = x_M \cos(\omega t + \phi)$. Here the circular frequency ω is fixed for given spring and given mass. It is related to the period of oscillations, T, by the well-known relation $\omega = \frac{2\pi}{T}$. The parameters x_M and ϕ characterize a particular oscillation mode. The maximal deviation x_M is called the *amplitude* of the oscillation and ϕ is called its *phase*.

Now it is not difficult to see that the oscillations[39] form a two-dimensional linear manifold. In other words, if $x_1(t)$ and $x_2(t)$ are any two oscillations, then $c_1 x_1 + c_2 x_2$ is also a possible oscillation. Moreover, if the two oscillations are

---

[39]They are the solutions of the Newton equation $a(t) = -\omega^2 x(t)$ where $a(t)$ is the acceleration of the bob. This equation is linear and thus the superposition principle is valid.

linearly independent,[40] then any oscillation may be expressed in this form by a proper choice of the parameters c_1, c_2.

To show this more formally, recall the simple trigonometric formula

$$\cos(\psi + \phi) = \cos \psi \cos \phi - \sin \psi \sin \phi.$$

Then obviously

$$x_M \cos(\omega t + \phi) = c_1 \cos(\omega t) + c_2 \sin(\omega t)$$

where $c_1 = x_M \cos \phi, c_2 = -x_M \sin \phi$.

Note that with linear problems, we need not even know the equation. It is sufficient to know the dimensionality, N, of the linear manifold of the solutions and to have N linearly independent solutions (they are said to form a *basis* of the manifold. Then we construct the general solution by using the superposition principle (writing arbitrary linear combinations).

Now you can sense why Einstein thought that Nature is nonlinear. Indeed, nothing can really happen in the linear world! In particular, there are no qualitative changes with changing environment or with parameters characterizing the system. These changes are typical features of the real world but are not present in simple linear models. For example, the period of the oscillations of the bob grows with its mass but nothing else changes. Of course, any real spring is not ideally elastic and it can be broken by a strong enough force. Thus one would expect nonlinear effects for heavy masses, and breaking of the spring would be their extreme manifestation.

To illustrate these statements let us solve a very simple nonlinear equation $y^2 + ax^2 = 0$. It is obvious that the variety of the solutions (x, y) strongly depends on the parameter a. If $a > 0$, the solution is just one point in the (x, y)-plane, $O = (0, 0)$. For $a = 0$ the solutions are points lying on the Ox axis, i.e., $(x_0, 0)$. Of course this is a linear manifold. If $a < 0$, then the solutions have the form $(x_0, x_0\sqrt{-a})$ or $(x_0, -x_0\sqrt{-a})$. Thus they lie either on the line OA or on the line OB (see Figure 3.9). As distinct from the case $a = 0$, the manifold of all solutions is now nonlinear. Indeed, the sum of the above solutions with the same x_0 is $(2x_0, 0)$ which is not a solution if $x_0 \neq 0$.

This example is very simple but sufficient to give a sense of the complexity of the nonlinear world. If you find that this illustration is too simple, try to find the manifold of the solutions of the equation $y^2 + ax^2 = bx$ depending on two parameters a and b. Even with very small b, this manifold is qualitatively different from the one considered above (corresponding to $b = 0$).

In "everyday physics" nonlinearity is a common thing. One may often see its characteristic manifestations — *strong dependence of parameters, creating or destructing new solutions.* For instance, when you try to push an unmoving car standing at rest on a flat place, you have to apply a stronger effort than later, when the car starts moving (it is better to have a companion to force the car to move but

---

[40]Linear dependence in this simple two-dimensional case means proportionality. More generally, the functions are linearly dependent if some linear combination of them is identically zero.

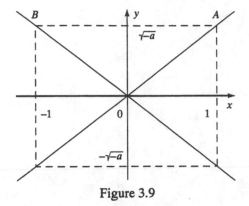

Figure 3.9

you can easily keep it rolling yourself, without aid). The car will not start to roll before the applied force exceeds a certain *threshold* value. This is because friction is essentially nonlinear — at rest it is significantly larger than in movement (the last is called rolling friction).

This sort of nonlinearity is accordingly known as a *threshold nonlinearity*. Below the threshold we have one state, immediately above the threshold, the system suddenly changes its state. In our example the car starts moving and can keep moving, for some time, even without any applied force.

The threshold type of nonlinearity is obvious in the nerve pulse excitation. Below the threshold the pulse is not excited at all. Without this threshold nonlinearity our life would be impossible and thus it is a fundamental thing. However, in the theory of solitons other, different sorts of nonlinearity play major roles. We discuss some of them in the following chapters.

Figure 9

Part II
Nonlinear Oscillations and Waves

There are many puzzles in the history of the soliton, but most mysterious is why the particular soliton to be discussed in this part of the book was not discovered in the 19th or in the beginning of the 20th century. Possibly, scientists encountered this soliton but did not realize its significance, although it has even more spectacular properties than other solitons.

For many years, physicists studied chains of pendulums and other similar systems modelling chains of atoms in crystals. Many sorts of travelling and standing waves in such systems were theoretically and experimentally studied. Thus, the physical properties of these systems were very well known, and numerous tools for their mathematical description were available. Nevertheless, as far as we know, nobody reported seeing solitons in such chains. Moreover, even after Ya. I. Frenkel and T.A. Kontorova theoretically predicted solitons in chains of atoms and studied their properties, their real significance as well as their relation to the Russell soliton remained unknown for almost 30 years.

This *Frenkel–Kontorova soliton* (we also call it the FK-soliton) is simpler than the Russell or KdV soliton. It is encountered in diverse physical systems and, as we shall see later, it is not very difficult to observe. The FK-soliton has a fixed shape that does not depend on its velocity. Curiously, the dependence of its energy E on the velocity v has the same form as for a massive relativistic particle

$$E = \frac{m_0 v^2}{\sqrt{1 - \frac{v^2}{v_0^2}}} \ ,$$

where m_0 is the mass of the particle. The only difference is that in place of the speed of light c, which is a universal constant of nature, there appears v_0 — the velocity of usual sinusoidal waves of small amplitude in the chain; this velocity depends on the parameters of the chain.

Thus, these solitons behave like relativistic particles. In addition, there also exist *antisolitons*, which are analogous to antiparticles. Solitons repulse other solitons but attract antisolitons. Moreover, soliton and antisoliton may form a bound particle-like system which is also a soliton (we may call it a "solitonic atom"). Most spectacular is that one may see all these phenomena in a very simple mechanical model that could easily have been constructed by Faraday, Maxwell, or Kelvin.

To understand the FK-soliton we must start from afar. First, we have to learn basic facts and concepts related to *nonlinear oscillations and waves*. A real understanding of the FK solitons requires some knowledge of the nonlinear pendulum and of waves in chains of bound pendulums.

Chapter 4
A Portrait of the Pendulum

And circular motion is more fundamental than rectilinear motion:
it is simpler and more perfect.

Aristotle

The Equation of the Pendulum

A pendulum is a small bob of mass m hanging on a rigid rod of length l. We will
study not only oscillations of the pendulum but also its rotations in a plane around
its suspension point. This means that the pendulum is suspended not from a point
O' but from an axle with a well lubricated bearing (see Figure 4.1). In fact, we
shall mostly treat the so-called *mathematical pendulum*, i.e., the bob is supposed
to be very small (mass point), the rod, absolutely rigid and very thin. Of course,
there must be no friction in the bearing.

To describe mathematically the movements of the pendulum, we introduce the
angle ϕ which is measured in radians and is supposed to be positive in the counter-
clockwise direction. Thus a full turn in the clockwise direction corresponds to ϕ
changing by -2π, a half-turn in the counter-clockwise direction corresponds to
a change of π, etc. A particular movement of the pendulum is then described
by a dependence of ϕ on time t, i.e., by a function $\phi(t)$. We may also use as a
coordinate of the bob the algebraic (positive or negative) value of the arc length
$s(t) = \phi(t)l$.

A.T. Filippov, *The Versatile Soliton*, Modern Birkhäuser Classics,
DOI 10.1007/978-0-8176-4974-6_4, © Springer Science+Business Media, LLC 2010

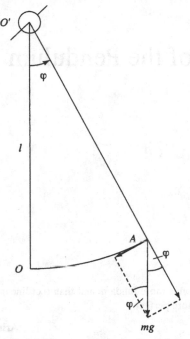

Figure 4.1 The mathematical pendulum

The movements of the bob can be determined from Newton's equation. To write it down we have to identify the forces acting on the bob. These are the vertical gravitational force mg and the tension in the rod's direction. The last one is neutralized by the component of the gravitational force $mg \cos \phi$, and thus the movements are completely determined by the tangential gravitational force $-mg \sin \phi$ (see Figure 4.1).

Now, the velocity of the bob, moving along the circle, is $v = s' = l\phi'$, where s' and ϕ' denote the derivatives in time. As we are interested only in tangential motions, we have to find only the tangential acceleration, which is $a = v' = s'' = l\phi''$ (here v'' and ϕ'' denote the second derivatives in time). Newton's second law equates the acceleration with the force and so

$$\phi'' = -\omega_0^2 \sin \phi , \quad \omega_0^2 = g/l . \tag{4.1}$$

This relation, expressing the angular acceleration of the bob $\phi''(t)$ in terms of its coordinate $\phi(t)$ at time t, is called the differential equation of its motion. To solve it is to find a dependence of the angle ϕ on time t such that the relation (4.1) is identically satisfied. The differential equation of motion describes all possible motions of the pendulum. To find any particular motion, we have to add some extra conditions. For example, if we know the initial position and the velocity of the bob, then the motion is completely determined by the differential equation.

Mathematicians would say that there exists a unique solution of the *initial value problem* $\phi(0) = \phi_0$, $\phi'(0) = \phi'_0$ (here ϕ_0 and ϕ'_0 are arbitrary numbers). The problem of finding the solutions of a differential equation with given initial values of the unknown functions is also called the *Cauchy problem*.

The equation (4.1) is *nonlinear*. In general, a linear combination, $c_1\phi_1 + c_2\phi_2$, of its two solutions ϕ_1 and ϕ_2 is not a solution. Even $c_1\phi_1$ is not a solution for $c_1 \neq 0$ because $\sin(c_1\phi_1) \neq c_1 \sin(\phi_1)$. Formally speaking, there exist some solutions ϕ_1 and ϕ_2, the sum of which is also a solution. Indeed, the obvious solutions $\phi = 0, \pm 2\pi, \pm 4\pi, \ldots$ and $\phi = \pm\pi, \pm 3\pi, \ldots$ satisfy this condition. The first series corresponds to the *stable equilibrium position* of the pendulum: the bob is at rest and its gravitational energy is minimal. The second series corresponds to the *unstable equilibrium position* of the pendulum: the bob is at rest but its gravitational energy is maximal. This means that even very small external influences (*perturbations*) can force it into motion.

Small Oscillations of the Pendulum

That was when I saw the Pendulum.

The sphere, hanging from a long wire set into the ceiling of the choir, swayed back and forth with isochronal majesty.

I knew — but anyone could have sensed it in the magic of that serene breathing — that the period was governed by the square root of the length of the wire and by π, that number which, however irrational to sublunar minds, through a higher rationality binds the circumference and diameter of all possible circles. The time it took the sphere to swing from end to end was determined by an arcane conspiracy between the most timeless of measures: the singularity of the point of suspension, the duality of the plane's dimensions, the triadic beginning of π, the secret quadratic nature of the root, and the unnumbered perfection of the circle itself.

Umberto Eco. Foucault's Pendulum

Ballantine Books, New York, 1989

The problem of describing all motions of the nonlinear pendulum is not easy to solve with the tools only from high school mathematics. However, if we are interested only in small amplitude oscillations, when $\phi \ll 1$, we may approximate $\sin\phi$ by ϕ and then the equations become approximately linear

$$\phi'' \approx -\omega_0^2\phi.$$

This is a simple example of *linearization*. One can now guess (or recall) a solution of this equation, $\phi = \phi_M \sin(\omega_0 t)$,[41] which vanishes for $t = 0$. Formally, the value of ϕ_M may be arbitrarily large, as the equation is now linear. However, we

---

[41] To check this one has to know the rules of differentiation of the trigonometric functions.

have to remember that, for large ϕ, the linearization is not valid. As $|\sin\phi| < 1$, ϕ_M must be small enough for the approximation $\sin\phi_M \approx \phi_M$ to be good.

The "sine" solution is only a special solution of the equation. There is another, independent solution, $\cos\omega_0 t$, which satisfies the initial condition $\phi(0) = 1$. Then, the most general solution of our linear equation can be written as a linear combination of these two independent solutions, i.e., by superimposing "sine" and "cosine" oscillations having arbitrary amplitudes.[42]

Now it is easy to guess that the most general oscillation is of the form $\phi = \phi_M \sin[\omega_0(t + t_0)]$ where the amplitude ϕ_M and the phase $\omega_0 t_0$ of the oscillation are arbitrary numbers. Using high school mathematics, one can see that this is equal to the linear combination of the sine-like and cosine-like oscillations, $\phi = c_1 \sin(\omega_0 t) + c_2 \cos(\omega_0 t)$ where $c_1 = \phi_M \cos(\omega_0 t_0)$, $c_2 = \phi_M \sin(\omega_0 t_0)$. Notice that c_2 is the initial position ϕ_0 and $c_1\omega_0$ is the initial angular velocity.

The solution of the equation for small oscillations may be found in a very simple way. We only have to recall the geometric definition of the trigonometric functions and the law of circular motion of the mass point. Let the point M move with a fixed velocity $V = \omega_0$ along the circle of the unit radius (see Figure 4.2). The velocity vector is tangential to the circle and its projection on the Oy axis is $\omega_0 \cos\alpha$ where $\alpha = \omega_0(t + t_0)$ radians (mind that t_0 is an arbitrary number). Then the point S, the projection of M, moves harmonically because $(OS) = \sin\alpha$ and the velocity of S is the projection of the velocity V on the Oy axis, i.e., $\omega_0 \cos\alpha$. Now, the acceleration a of the point M is directed to the center O and is equal to $V^2 = \omega_0^2$ (recall that the radius of the circle is unity). The acceleration of the point S is the projection of a and thus it is $-\omega_0^2 \sin\alpha = -\omega_0^2(OS)$. Denoting (OS) by ϕ, we have the equality $\phi'' = -\omega_0^2\phi$, which is just our equation, if we set $\omega_0^2 = g/l$.

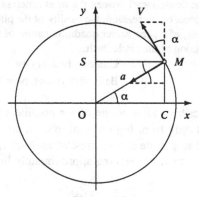

Figure 4.2

---

[42]We need not bother about making these amplitudes small. Due to the linearity of the equation, any linear combination will be a solution, but only those having small amplitudes will approximate the original nonlinear problem.

Galileo Galilei's Pendulum

Simplicio: "Concerning natural things we need not always seek the necessity of mathematical demonstrations."

Sagredo: "Of course, when you cannot reach it. But if you can, why not?"

Dialogue on the two Major Systems of the World.
Galileo Galilei (1514–1642)

From the above, it must be clear that the period of the linear oscillations is

$$T = \frac{2\pi}{\omega_0} = 2\pi\sqrt{\frac{l}{g}}.$$

This well-known formula was discovered by Huygens.[43] The essence of this formula, which is the square root dependence of T on l, was earlier known to Galileo, although he had not written it explicitly. Some people believe that Galileo found it experimentally, but this is not quite true. In experiments, he noticed a dependence of the period on the length of the pendulum and then discovered the proportionality of the period to the square root of the length by very beautiful arguments.

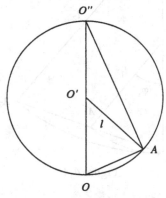

Figure 4.3

In fact, the most important observation he made about the pendulum was independence of the period on the amplitude. This he formulated as the equality of the periods for pendulums of equal lengths ("isochronism"). Then he used the law of free fall which he had discovered earlier and also a relation between the slip of a body along an inclined plane and its free fall from the same height.

---

[43] He first discovered independence of the period on the amplitude (*isochronism*) for the oscillations of the cycloidal pendulum (the mass point moving along the cycloid). Using the concept above, he found the acceleration of the mass point moving along the circle and the formula for the period of the small amplitude oscillations.

Let us first give Galileo's slightly modernized s arguments.[44] First, we approximate the motion of the mass point along the arc AO by its sliding down the chord AO (Figure 4.3). Then, the time t of this sliding is equal to that of the free fall from O''. This follows from the fact (known to Galileo) that the ratio of the acceleration along AO to g (the acceleration along $O''O \equiv 2l$) is equal to the ratio $AO/O''O$. As $2l = \frac{1}{2}gt^2$, we find that the quarter period of the oscillation must be $2\sqrt{l/g}$, i.e., the period is $T = 8\sqrt{l/g}$. The incorrect factor 8 instead of 2π is, of course, the result of approximating motion along the arc by sliding down the chord.

Galileo's original reasoning is easy to understand after looking at Figure 4.4. Let us consider two pendulums attached to O_1 and O_2, respectively. The time of sliding down the inclined plane A_2A_1O is proportional to the square root of OA_1 for the first pendulum, and to the square root of OA_2 for the second pendulum. The lengths of OA_1 and of OA_2 are respectively proportional to the lengths of the pendulums $(OO_1) = l_1$ and $OO_2 = l_2$. Recalling now the law of free fall, we find that for *similar oscillations* (i.e., with the same angular amplitude ϕ_M) the periods must be proportional to the square roots of the lengths. Due to isochronism this must be true for any oscillations.

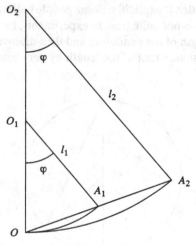

Figure 4.4

For small oscillations, Galileo's arguments are completely correct. It is easy to understand that *the isochronism is a very simple consequence of the linearity of small oscillations*. Indeed, if we know a solution $\phi_M \sin[\omega_0(t+t_0)]$, then any other solution is obtained simply by changing the amplitude ϕ_M and t_0 (the frequency, ω_0, is fixed). Thus the period is the same for all solutions.

A very interesting feature of Galileo's argument is that it hinted at a very general idea about similar behavior of similar systems. In a more explicit form, the general *similarity (similitude) principle* was first formulated by Newton and

---

[44] In this we follow the beautiful *Lectures on Oscillations* given by L.I. Mandelshtam at the Physics Faculty of the Moscow University in the 1920s.

Hooke. Later, this principle gave birth to a very useful and simple *dimensional analysis*. By using ABC of this analysis, the Huygens formula may easily be found (apart from the factor 2π) without computations or experiments. So, let us briefly digress on this principle and its simplest applications.

About Similarity and Dimensions

Small is similar to great,
Though they different appear.

Goethe (1749–1832)

Newton–Hooke's principle of similarity was forgotten for about one hundred years before Fourier restored it to life and further developed it in the above-mentioned *Analytical Theory of Heat* in which he introduced the notion of *dimensionality* of physical variables and the *principle of homogeneity in dimensions*.

What is dimensionality? To express measurable mechanical quantities in terms of numbers, we have to introduce units for three basic quantities. For these, physicists usually take the units of length L, of time T, and of mass M. Other quantities are *derived (derivative)* ones; this means that the units for them can be derived from the three basic units. For instance, we need not introduce a unit of area, because the area of any rectangle S is simply the product of the lengths of its sides (the area of any figure can be measured by dividing it into small rectangles). If the lengths of the sides are multiplied by the same number c, the area will be multiplied by c^2. This means that the dimensionality of the area is equal to the dimensionality of length squared. This fact we express by the *dimension formula* $[S] = L^2$.

The dimension formula says that the area of any figure is multiplied by c^2 if its linear dimensions are c times larger (as in magnifying a photo). If we did not know how to calculate the area of a figure, the dimension formula would help us to relate the area to its linear dimensions. For example, the area of a circle of radius R must be $S = aR^2$, where a is some number that cannot be determined by pure dimensional considerations. However, we know that this number is universal, i.e., the same for all circles. Thus, by measuring the area of one circle we can find this number (approximately), and so will know the area of all circles. By similar dimensional arguments, one can find that the volume of the sphere of radius R must be $V = bR^3$, while the area of the sphere must be cR^2.

Starting from the definition of the velocity of uniform motion, $v = (x_1 - x_2)/(t_1 - t_2)$, one may write the dimension formula for the velocity $[v] = LT^{-1}$. Its meaning is simple: when all distances are magnified by a factor of c_L while the time intervals by a factor of c_T, the velocity is multiplied by the factor c_L/c_T. Usually, when a physical variable is introduced, one indicates also units in which it is measured (e.g., $S[cm^2]$, $v[cm\ sec^{-1}]$, etc.). This, of course, gives the dimension of this variable. Thus, once we know that the acceleration is measured in $cm\ sec^{-2}$ we have the dimension formula $[a] = LT^{-2}$. The dimension formula is a more general thing than the indication of units; it is not changed if we use meters instead of centimeters and hours instead of seconds, etc.

Above, we have mentioned kinematical variables, but the dimensional analysis is equally applicable to dynamical variables, which depend also on the unit of mass. For instance, the dimension formula for the force variable is $[F] = MLT^{-2}$, for the energy $[E] = ML^2T^{-2}$, and so on. The exponents in the dimension formulae are called the dimension indices. We may operate with them as with usual algebraic exponents.

Let us consider, for the sake of illustration, Newton's formula *Force = Mass × Acceleration*. This definition gives the dimension formula $[F] = MLT^{-2}$. The same dimension formula may be written $[F] = M[a]$. More generally, one may work with the dimensional formulae exactly as with usual algebraic ones.

The principle of uniformity in dimensions states that both sides of any equation relating physical variables must have identical dimension formulae, i.e., equal dimension indices. This rule is well-known, and everybody uses it to check calculations. If a result of some calculation gives wrong dimensionality for some variable, we have to look for an algebraic mistake. More interesting is another way of using this principle, that is, for deriving new formulae relating physical variables.

Let us demonstrate this by deriving Galileo's law for free falling bodies. It is obvious that the path s may depend on time t, on mass m, and on the force, which is mg. So, we may write $s = kt^d m^b (mg)^c$, where the dimension indices are to be found from dimensional considerations, while k is a dimensionless constant (a number). The dimension formula for the right-hand side is $T^d M^{b+c}[a^c] = M^{b+c}T^{d-2c}L^c$. This must be equal to the dimension formula of the left-hand side which is simply $[s] = L$. Equating the indices we have $c = 1$, $d - 2c = 0$, $b + c = 0$. It follows that $s = kgt^2$, where the dimensionless number k cannot be found from the dimensional consideration (we know that $k = 1/2$).

Exercise. Using the dimension formula for acceleration, find the acceleration of a uniformly rotating mass point (centripetal acceleration) in terms of the angular velocity and circle radius. Find by dimensional considerations its kinetic energy.

Quite similarly, we may derive Huygens' formula for the period of linear oscillations of the pendulum. The period T may depend on the length l, mass of the bob m, and on the force f that is applied to it. Thus, we find that $T = dm^a f^b l^c$ and the dimension equation is $[T] = M^{a+b}L^{b+c}T^{-2b}$. It follows that $a + b = 0$, $b + c = 0$, $-2b = 1$ and the formula for the period is

$$T = d\sqrt{ml/f}.$$

With $f = mg$ we have the Huygens formula up to the undetermined constant d (Galileo would have found an approximate value $d \approx 8$ and Huygens had obtained the correct result $d = 2\pi$).

It is of interest to note that, in this formula, f need not be the gravity force. For example, let the bob bear the electric charge q and the pendulum be placed in a uniform electric field

E, say between faces of a big charged capacitor (strictly speaking, between two infinite charged planes). Then we have to substitute in our formula the expression $f = mg + qE$. We know that for $E = 0$, the value of d is 2π. As the number d is independent of E, we find the exact formula for the period of oscillations in gravitational and electric fields

$$T = 2\pi \sqrt{\frac{ml}{(mg + qE)}}.$$

Of course, such extremely simple considerations do not necessarily lead to complete formulae. For example, let us try to find the period of nonlinear oscillations of the pendulum. The linearity in the above reasoning told us that the period does not depend on the amplitude, but now this is not true. However, after some reflection, you may convince yourself that our dimensional considerations are also valid for nonlinear oscillations, but d must depend on the amplitude through some dimensionless combination, e.g., the ratio of $s_M = l\phi_M$ to l. Thus the period of arbitrary oscillations of the pendulum must be

$$T = 2\pi d(\phi_M)\sqrt{\frac{l}{g}}. \tag{4.2}$$

Recalling that $d(\phi_M) \approx 1$ for small values of ϕ_M, we conclude that $d(0) = 1$.

It is also not difficult to show that $d(\phi_M) > 1$. This follows from the inequality $|\sin \phi_M| < \phi_M$. The physical meaning of this inequality is that the force returning the pendulum to its equilibrium position is smaller for nonlinear oscillations. This results in the smaller acceleration and thus in the larger period. Thus the period is growing with ϕ_M, and one may show that $d(\phi_M)$ infinitely grows when ϕ_M approaches π (the oscillations exist only for $\phi_M < \pi$, otherwise the pendulum will rotate).

We see that very simple arguments allow us to learn many interesting and not obvious properties of the pendulum, which, by the way, is not a very simple physical system.

I need not spend much time explaining why an apparently simple device known as the pendulum is in fact a rather complex mechanical system. You must feel it by now. Of course, it was not accidental that Galileo, Huygens, and Newton used it as much as they did in discovering the laws of mechanics. Even today the pendulum is used as a generic example of a nonlinear system in any textbook on nonlinear problems, and a significant part of the physicist's intuition on possible effects of nonlinearity has its origins in the behavior of pendulums.

I wish also to say that there are very good reasons to start a deeper study of solitons with some lessons about the pendulum. You will soon see that there is a deep relation between mathematical laws describing the pendulum and those describing solitons. Moreover, some simple and intuitively clear examples of solitons may be physically realized by using mechanical devices made of the pendulums or electrical devices mathematically equivalent to systems of the pendulums.

Returning to dimensional analysis, I have to caution againt overestimating the simplicity of dimensional analysis and the similarity principle. Quite often, this simplicity is delusional! In fact, to effectively use these powerful tools, a deep enough understanding is required of the essence of the physics problems involved. Unfortunately, it is difficult to formulate what "deep enough understanding" means so that it is difficult to give general formal rules for applying these beautiful things to diverse concrete problems.[45] While dimensional analysis gives the answers really very fast in some simple problems, its application to more complex ones can be an art form rather than a science. For this reason, Newton's great discovery was slow in becoming generally recognized, and its applications raised numerous debates. About 90 years ago, Sir John William Rayleigh lamented over indifference of his contemporaries to the "Great Principle of Similarity and Dimensional Analysis." He rightly complained that often well-known and respectable scientists were spending a great deal of effort to discover "new laws" which could be derived very quickly by using dimensional analysis.

I hope that now you are interested enough to try and apply this principle to discovering new formulae. You will find other interesting examples in the following chapters.

Energy Conservation

You were caught on the swing,
Be swinging, devil with you!

Vladimir Sollogub (1863–1927)

The equation of motion gives a complete description of the pendulum. An alternative description, which is more appealing to our intuition, is based on using the energy conservation principle. This approach is especially convenient and effective when dealing with nonlinear problems.

The kinetic energy of the pendulum is $T = \frac{1}{2}mv^2 = \frac{1}{2}m(\phi')^2$. The potential energy is easy to derive from Figure 4.5, and is equal to $mg(OH)$, where $(OH) = (OA)\sin(\phi/2) = 2l\sin^2(\phi/2)$. Thus the potential energy is $U = 2mgl\sin^2(\phi/2)$ and the full energy (kinetic plus potential) is

$$E = \frac{1}{2}ml^2 T(\phi')^2 + 2mgl\sin^2(\frac{\phi}{2}) .$$

If there are no friction forces, this energy is constant, i.e., independent of time. Then this equation expresses the law of energy conservation for the pendulum, and it is valid for arbitrary motions of the pendulum.

---

[45]Try for instance to understand why the principle of similarity is apparently not applicable to living creatures. Can there exist a giant man whose linear dimensions are ten times larger than the dimensions of "normal" men?

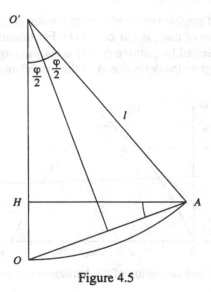

Figure 4.5

It is more convenient to rewrite this equation in a dimensionless form:

$$\frac{(\phi')^2}{\omega_0^2} + 4\sin^2(\frac{\phi}{2}) = \frac{E}{m\omega_0^2 l^2/2} \ . \tag{4.3}$$

The right-hand side is the ratio of the energy of the pendulum E to the kinetic energy of the massive point that is uniformly rotating along the circle of radius l with the period $T = 2\pi/\omega_0 = 2\pi\sqrt{l/g}$. Denoting this energy by E_0, we may write the right-hand side as E/E_0. Then the first term in the left-hand side is the dimensionless kinetic energy T/E_0, while the second term is the dimensionless potential energy U/E_0.

When the amplitude of the oscillations ϕ_M is small, we may approximate $\sin(\phi/2)$ by $\phi/2$. Then the law of energy conservation (4.3) looks simpler

$$\frac{(\phi')^2}{\omega_0^2} + \phi^2 \approx \frac{E}{E_0}, \quad E_0 \equiv \frac{1}{2}m\omega_0^2 l^2 \ . \tag{4.4}$$

The right-hand side of (4.3) is convenient to rewrite in terms of the amplitude ϕ_M. In the apex position, where $\phi = \phi_M$, the pendulum angular velocity is zero and thus the energy conservation (4.3) gives

$$\frac{2E}{m\omega_0^2 l^2} \equiv \frac{E}{E_0} = 4\sin^2(\frac{\phi_M}{2}) \approx \phi_M^2 \ ,$$

where the approximate equality refers to the small amplitude, linear oscillations.

The energy conservation equation allows us to introduce more intuitive ways for describing the dynamics of the pendulum as well as of other more complex

systems. The standard graphic representation of motion is by drawing the dependence of the coordinate of time, in our case $\phi(t)$. For example, simple harmonic
motion may be represented by plotting $\phi = \phi_M \sin(\omega_0 t)$ against t, as in Figure
4.6. This plot simply gives the deflection ϕ of the pendulum at any moment t.

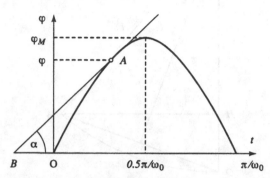

Figure 4.6 Graphical representation of the harmonic motion $\phi = \phi_M \sin(\omega_0 t)$

Using this graph, one can also find the velocity, but this is not so simple. The
angular velocity ϕ' is defined by the angle α between the tangent BA to the curve
$\phi(t)$ at the point $(t, \phi(t))$ and the t-axis, in fact, $\phi' = \tan \alpha$. So the graph does not
directly show the velocity. It is even more difficult to find the potential or kinetic
energy from this graph.

Let us draw the dependence of the potential energy $U(\phi)/E_0$ on the position
(angle) ϕ. In Figure 4.7b, OA represents $\phi(t)$, AA_2 is E/E_0, $AA_1 = U/E_0$,
and $(A_1A_2) = T/E_0$. As the full energy E is constant, the point A_2 is moving
along the straight line A_2M while A_1 is moving along the parabola OM defined
by the equation[46] $U(\phi)/E_0 = \phi^2$. This picture is called the *potential diagram* of
the pendulum. It is more informative than the simple graph of motion in Figure
4.6, but it is useful to combine both of them as shown in Figure 4.7. Looking at
diagrams 4.7a and 4.7b, we may find very complete information about the motion
— position, kinetic and potential energy as well as amplitude and the full energy
can be read immediately. The only inconvenience is that the square of the velocity,
rather than the velocity itself, is directly represented (by A_1A_2).

To follow simultaneously both the position and velocity, consider one more
diagram, Fig.4.7b, in which we draw the dependence of ϕ'/ω_0 on ϕ obtained
by solving the equation (4.4). So, when the bob is moving from the equilibrium
position $\phi = 0$ to the apex $\phi = \phi_M$, the point A_3 moves along the circle of radius
$\phi_M = \sqrt{E/E_0}$. This is easy to see from (4.4), as $(OA) = \phi(t)$ and $(AA_3) =
\phi'(t)/\omega_0$. For the simple harmonic motion we have $(OA) = \phi_M \sin(\omega_0 t)$ and
$(AA_3) = \phi_M \cos(\omega_0)$. This obviously means that the point A_3 moves along the
circle uniformly, with constant velocity.

---

[46]We consider first the linear approximation and accordingly the energy conservation law (4.4).

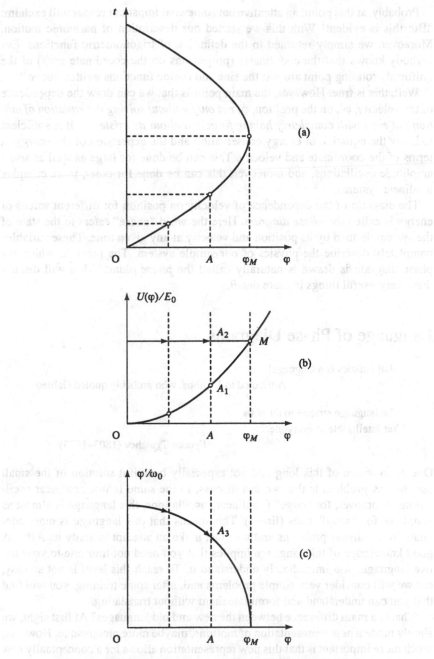

Figure 4.7 Different representations of harmonic motion:
a) graph, b) potential diagram, c) phase diagram

Probably, at this point, an attentive but somewhat impatient reader will exclaim: "But this is evident! With this we started our description of harmonic motion. Moreover, we simply returned to the definition of trigonometric functions. Everybody knows that the coordinates (projections on the coordinate axes) of the uniformly rotating point are just the sine and cosine functions written above."

Well, this is true. However, the main point is that we can draw the dependence of the velocity, ϕ', on the position, ϕ, *not only without solving the equation of motion but even with completely having forgotten about its existence*. It is sufficient to know the equation of energy conservation and the expression of the energy in terms of the coordinate and velocity. This can be done for large as well as small amplitude oscillations, and moreover, this can be done for other, more complex nonlinear systems.

The diagram of the dependence of velocity on position for different values of energy is called the *phase diagram*. Here the word "phase" refers to the state of the system defined by its position and velocity at any given time. These variables completely describe the physics of our simple system. The plane in which the phase diagram is drawn is naturally called the *phase plane*.[47] We will discuss these very useful things in more detail.

Language of Phase Diagrams

Mathematics *is* a language!
> Attributed to J. Gibbs, who probably quoted Galileo

The language strange to all of us
Yet intelligible to everyone...
> Fyodor Tyutchev (1803–1873)

One main virtue of this long and not especially beautiful solution of the small oscillations problem is that we can discuss, in the same terms, nonlinear oscillations. Moreover, for "large" (nonlinear) oscillations, this language is almost as simple as for "small" ones (linear). This means that this language is more adequate for nonlinear problems and we must make an attempt to study its ABC. A good knowledge of this language implies that you need not translate to your native language, you immediately understand it. To reach this level is not so easy, but we will consider very simple problems and, after some training, you will find that you can understand and formulate them without translating.

What is a main difference between the new and old languages? At first sight, we simply made a new representation of motions, maybe more picturesque. However, much more important is that this new representation allows for a conceptually new approach to the problem. The crucial point is that to draw the phase diagram, we

---

[47]For more general systems, described by many variables, one speaks of corresponding phase spaces. All motions of a physical system can be depicted by curves in its phase space.

need not solve any differential equation! Instead, the most important qualitative information about the motions of a system may be extracted by viewing its *phase trajectories* in the diagram. So we call the curves in the plane $(\phi, \phi'/\omega_0)$ corresponding to different values of the energy. By looking at phase trajectories it is easy to find maximal amplitudes, maximal velocities, and so on.

It is also easy to grasp a qualitative picture of the motions of a pendulum. In fact, to understand them, we do not have to know the exact position of the pendulum at any given time. The qualitative picture, represented by the phase diagram, is really more important! Moreover, any particular motion may be found from its phase trajectory, which gives the dependence of ϕ' on ϕ for a fixed value of the energy. This is an easy problem for personal computers.

The language of phase diagrams and phase trajectories has been used systematically in nonlinear scientific problems for about 70 years, but it was invented more than a century ago. It requires first of all a very good understanding of the energy conservation law. Its mathematical expression for general mechanical problems was made particularly clear by the German mathematician Karl Weierstrass (1815–1897). He regarded the relation (4.3) as an equation for the function $\phi(t)$. To describe its solutions he used elliptic functions.[48] After Abel and Jacobi, he made a major contribution to the theory of these functions. This theory and its deep generalizations still attract mathematicians. Physicists and engineers apply these functions to many nonlinear problems. Note also that elliptic functions play an important role in soliton theory. Unfortunately, we must pass over these beautiful mathematical buildings. Fortunately, the main features of many interesting physical systems may be understood using the much simpler language of phase diagrams.

To my knowledge, such an application of phase diagrams was first made in 1885 by the French mathematician Henri Leoté. He was an instructor of mathematics at *l'École Polytechnique*[49] (professor from 1895). He specialized in mechanics and used phase diagrams to analyze some automatic controllers (a prototype was James Watt's regulator).

Leoté did not attempt to construct a general mathematical theory and his approach to phase diagrams was rather physical. He was not aware that, at the same time, a mathematical foundation of a more general theory had been established. In fact, in 1885, a young French mathematician Henri Poincaré (1854–1912) published a paper "On curves defined by differential equations," in which he summarized a conceptually new approach to studying solutions of nonlinear differential equations (he developed his views in 1879–1885).

First, Poincaré's concept of what it means to solve a differential equation was radically different from the standard one. The main idea was to study the func-

---

[48] The simplest elliptic functions introduced by Jacobi generalize trigonometric functions.

[49] This is a most important French technical university, having had such students as Ampère, Arago, Poisson, Cauchy and other celebrated scientists. Leoté and Poincaré also graduated from this institute.

tions defined by differential equations (their solutions) without really solving the equations in analytic form. He expressed his views very clearly.

> A complete study of a function consists of two parts: 1) a qualitative, or geometric analysis of the curve defined by the function; 2) a quantitative, or numerical evaluation of the function. . . . Thus, to study an algebraic function, one begins by drawing the curve defined by this function. . . i.e., one finds closed and infinite branches of the curve, etc. After this qualitative analysis one may numerically find a certain number of particular points of the curve.
> It is natural to begin a theory of any function with a qualitative approach. For this reason, the first problem in solving a differential equation is to draw the curves defined by it. This is a qualitative investigation. Being completed, it enormously simplifies the numerical analysis of the solutions. . . . Besides, this qualitative study is of primary interest by itself. Diverse and very important problems in mechanics and analysis may be reduced to it.

Nowadays, these views are quite natural, almost self evident, but a century ago this was not so. At that time such concepts seemed very unusual, if not eccentric, and were difficult to understand and almost impossible to accept. In addition the intuitive, geometric style of Poincaré's works was at variance with the canons of rigorous mathematical deduction of Cauchy and Weierstrass which had been accepted by a majority of mathematicians. Today, the standards of mathematical rigor are even more severe but, at the same time, many more technical tools for making rigorous proofs in qualitative analysis are available. Poincaré was the creator of a new field in mathematical physics and discovered a world of new things. Naturally, it was not a priority for him to be completely rigorous. As a matter of fact, many of Poincaré's statements were not rigorously proved by him and some of them (not many) were later found to be false or incorrectly formulated. However, the plan of the whole building of qualitative analysis was completely correct.

Gradually, as Poincaré became one of the most famous mathematicians of the world,[50] his works attracted an increasing amount of attention. So, about 25–30 years after the "Curves" were published, a rigorous basis for his qualitative theory of differential equations was established. Still today, this is a living theory and its generalizations are applicable not only to differential equations (these generalizations are generically called dynamical theory, or simply "dynamics"). You can find references to methods and results of Poincaré in any book on nonlinear oscillations, nonlinear equations or general dynamical theory. His ideas were fundamental for all these modern theories.

The fate of Leoté's work was different. Neither Leoté nor other scientists, including Poincaré, noticed a relation between Leoté's phase diagrams and Poincaré's general theory, and his paper was completely forgotten. Other works of

---

[50]Like Gauss or Euler, he worked in almost all branches of mathematics and physics of his time. As a professor at the Sorbonne, he gave lectures on a new subject every year from 1881 until his early death!

his on the theory of machines and mechanisms (a branch of applied mechanics) were highly respected, and in due time (in 1890) he became a full member of the *Académie de Science de Paris*. However, his most original work was forgotten. It was discovered after almost 50 years of oblivion by the Russian physicist Alexander Andronov (1901–1952).

Andronov was one of the bright stars of the Russian scientific school of Leonid Isaakovich Mandelshtam (1879–1944). Under Mandelshtam's influence, he began studying nonlinear oscillations. He "discovered" the works of Poincaré while a postgraduate under Mandelshtam. Soon he realized that the mathematical language created by Poincaré was a most effective tool for solving problems of nonlinear oscillations. Mandelshtam found this idea relevant and very attractive.

The result was that a new branch of science was born, later successfully developed by Andronov and his pupils, who made important contributions to mechanics, physics and radio engineering. [51] In addition, they initiated interesting new developments in mathematics. I note an influential paper by Andronov and L.S. Pontryagin,[52] in which the concept of "robust systems" was introduced. Roughly speaking, this means that the phase diagrams do not qualitatively change under any small enough perturbations. This is a good example of interrelations between physics and mathematics. Poincaré once said about the usefulness of these interrelations:

> Physics can't exist without mathematics which provides it with the only language in which it can speak. Thus, services are continuously exchanged between pure mathematical analysis and physics. It is really remarkable that among works of analysis most useful for physics were those cultivated for their own beauty. In exchange, physics, exposing new problems, is as useful for mathematicians as it is a model for an artist.

This very clearly describes the relations between physics and mathematics in general and between oscillation theory and differential equations in particular. However, you see that this statement was made by a mathematician. Physicists will never agree even to compare their science with a model for mathematics. Instead they usually talk about mathematical models for physical phenomena. In fact, the most important part of the physicist's work is to find a mathematical

---

[51]I wish to specially mention young and bright A.A. Witt. He died in Stalin's concentration camps but this was not enough since even his name disappeared from the title of the fundamental book *Theory of Oscillations* by Andronov, Witt and S.E. Khaikin. His name was restored in the second edition published after Stalin's death.

[52]The best known achievements of Lev Semenovich Pontryagin (1908–1988) were in topology but his interests were very wide and for many years he worked in the theory of differential equations with applications to the theory of automatic control. In 1955 or 1956, as a student of the physics faculty, I attended his lectures on differential equations and his seminar on a general approach to automatic control at the mathematics faculty. His lectures as well as all his books were extremely lucid. Possibly, this extremely condensed and clear style was developed from his childhood. At the age of 14, he completely lost his vision in an accident with primus.

model describing principal features of a phenomenon under study. The next stage — mathematical investigation of the model — is closer to the work of the "pure" mathematician. But even here, the physicist's approach differs from the mathematician's. The physicist finds inspiration not in mathematical beauty but rather in physical intuition. The very concept of what it means "to solve a problem" is different for the physicist and the mathematician. Thus Poincaré's aphorism is "truth, only truth, nothing except truth" but certainly not "all truth." A "pure" physicist L.I. Mandelshtam treated the relation between physics and mathematics in the theory of oscillations as follows.

> Well, once you deal with equations, mainly with differential equations, you are doing, in some sense, mathematics. But this is not the point. First of all, because it is physics that teaches us how to "interrogate" the differential equations. In the theory of oscillations, mathematical patterns... have a very clear intuitive content, not only geometrical but also physical. In other words, the analysis here is supported both by geometric and physical intuition... this intuition may be very rich and broad, having the roots in radio engineering, electrical engineering, optics, and so on.

Digression on Celestial Mechanics, Perturbations, and Stability

Above, we emphasized qualitative methods for studying nonlinear oscillations. But a purely qualitative approach cannot give a complete solution of a physical or engineering problem. Physicists, engineers, astronomers need precise numerical relations between measurable quantities. For all applications as well as for further theoretical treatment of a physical problem, an *analytical* solution, expressed in terms of known functions and standard mathematical operations is indispensable.

The development of general analytical methods for deriving motions of complex systems started in the 18th century. The challenge was to calculate the orbits of the planets, which is not a problem at all if one neglects the attraction between the planets and only takes into account their attraction to the sun. However, if you wish, say to make a precise prediction of the position of the moon, you have to take into account forces acting between at least three objects — the sun, the earth, and the moon. This is the famous "three-body problem" which is not yet completely solved.

First attempts to solve the above problems were made by D'Alembert and Euler. They proposed the perturbation methods idea, which says essentially that the strongest forces defining the main features of motions are taken into account exactly, while the effects of smaller forces (called perturbations) are calculated only approximately. This general approach is called *perturbation theory*.

Laplace and Poisson developed D'Alembert's and Euler's idea in more detail. In particular, Poisson noted that the idea may be applied to computing oscillations of the nonlinear pendulum. In this case, the unperturbed system is simply the linear pendulum ($\sin \phi$ is approximated by ϕ) and the perturbation is defined by small nonlinear corrections ($|\sin \phi - \phi|$ is small for small ϕ). Poisson's method gives good results if ϕ is small enough and if the time interval for the study of motion is not too large. Note that Ostrogradskii later improved Poisson's method and obtained the perturbed solutions for infinitely large time intervals.

The general scheme of perturbation theory in celestial mechanics was very similar, although the computations were much more complex. Lagrange and especially Laplace performed extensive and laborious computations of perturbed motions (the unperturbed ones are the movements along Kepler's orbits). Their results allowed them to find the exact positions of the planets in the remote past and in the future. Using their methods, Adams and Le Verrier found a deviation of the real orbit of the planet Uranus from the computed one. To explain this deviation they boldly suggested that it is due to a new perturbation caused by an unknown planet (Neptune).

Poincaré and the Russian mathematician Aleksandr Mikhailovich Lyapunov (1857–1918) improved and generalized perturbation methods. They also had in mind applications to celestial mechanics, but their methods were so general that applications to solving seemingly very different nonlinear problems in physics and technique appeared as soon as these problems were properly formulated. When Mandelshtam and Andronov began applying the methods of Poincaré and Lyapunov to nonlinear problems in radiophysics they were astonished how nicely these methods fit into this new field.[53]

Around the same time, Nikolai Mitrofanovich Krylov (1879–1955) and 17-year-old Nikolai Nikolaevich Bogoliubov (1909–1993) were developing new perturbation methods, which allow one to compute both periodic and much more complex non-periodic motions of nonlinear systems. Later, Bogoliubov further generalized these methods to make them applicable to chaotic motions in systems consisting of an enormously large number of particles. Thus he succeeded in justification of some of the hypotheses in statistical mechanics of Boltzmann and Gibbs.

Later, especially under the influence of ideas of Andrei Nikolaevich Kolmogorov (1903–1987) and of his student V.I. Arnold (now a renowned mathematician himself), a unification began between quantitative (perturbative) and qualitative (topological) methods. This resulted in the flourishing of nonlinear mechanics. The methods used were thereafter applied to other sciences and play a prominent role in soliton theory. Note that this general science is now usually called the theory of dynamical systems, or abstract dynamics.

---

[53]The system of planets does not look very much like a radio transmitter but the perturbation methods are equally applicable to both.

Phase Portraits

Let us return to studying pendulum motions in a manner suggested by our physical and, in part, geometric intuition. We may draw the phase trajectories for any energy of the pendulum, thus embracing both periodic (oscillatory) and aperiodic (rotatory) motions. The totality of all possible phase trajectories form the phase portrait of the pendulum, which visualizes these motions in a clear manner.

To learn how to draw and understand phase portraits, let us first look at the simplest problems. Suppose that a point particle moves with a constant velocity v_0 along a straight line. Choosing this line as the coordinate x-axis, we may easily draw the graph of this motion, e.g., as shown in Figure 4.8a. This graph may look slightly unusual because we have chosen time t as the vertical axis, considering dependence of time on the path, $t = (x - x_0)/v_0$, instead of the usual graph of $x - x_0 = v_0 t$.

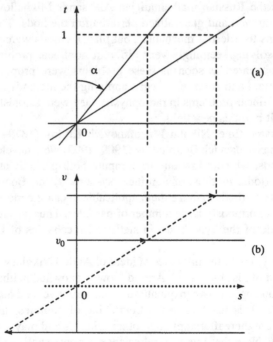

Figure 4.8 Uniform motion: a) graphs of motion with $s_0 = 0$, b) phase diagram

If, on the vertical axis, $1cm$ corresponds to $1sec$ and, on the horizontal axis, $1cm$ corresponds to $1cm$ of the path, then the velocity is equal to $\tan(\alpha)$ cm/sec. In what follows we will not repeat these conventions for other physical variables having other dimensions. Thus, it must be clear that on the v-axis (v is the velocity) $1cm$ corresponds to the velocity $1cm/sec$. By the way, negative values of α correspond to motions in the negative direction of the x-axis.

It is easy to draw any phase trajectory: it is defined by the value of the velocity v_0, Figure 4.8b. Phase trajectories are simply the straight lines parallel to the x-axis crossing the v-axis at $v = v_0$. For positive velocity, the point on the phase diagram, representing the material point, moves in the positive direction; for negative velocity, in the negative direction. Note that while the graph of the motion depends on the initial position x_0, the phase trajectory is independent of it.[54] Finally, if the point is at rest ($v_0 = 0$), the graph is the line parallel to the time axis, i.e., $s = s_0$, $\alpha = 0$. The image of this "motion" on the phase diagram is the point x_0 on the x-axis, namely the point $(x_0, 0)$. For different values of s_0, these points fill out the whole axis (each point of the x-axis is a separate phase trajectory). Thus, the phase trajectories of the point uniformly moving along the x-axis are straight lines parallel to it and also the points of the x-axis. The important fact is that only one trajectory passes through each point of the phase plane (x, v). This statement is true if we neglect t-dependence, i.e., if we identify trajectories with different x_0 (or t_0); we may say also that the phase trajectory is a directed but not parametrized curve.

It is a good exercise to draw the phase diagram for movements of a small mass, falling from some height or thrown up vertically. Let me stress that we are considering only precisely vertical motions, (imagine that the mass moves in a thin tube). The point O on the phase diagram will correspond to the mass at rest in the lowermost position (on the surface of the earth). This point is therefore called the *equilibrium point*.[55] The mass can stay at this equilibrium point for an indefinitely long time. In addition, if we slightly push the mass at this point, it will oscillate near this point; due to friction, it will eventually come to rest at O. So, this is a *stable equilibrium point*.

The equilibrium points on the phase portrait of the uniformly moving mass (i.e., the points $(x_0, 0)$) are *unstable equilibrium points*. If we slightly push our mass, it starts moving away from the initial equilibrium position. On the phase portrait, the point "jumps" to a nearby phase trajectory and will remain on it to infinity. Of course, in any real physical system (say, a hockey puck on ice) there is some sort of friction, which may significantly change the whole portrait. I suggest you draw the phase portraits of the puck for very small amounts of friction (on good ice) and for extremely large amounts of friction (say, on a dry asphalt pavement).

Phase Portrait of the Pendulum

> You will never find two completely identical creatures in Nature.
>
> Ludwig Boltzmann

---

[54] In other words, the phase trajectory does not depend on the moment t_0 in which we start counting time.

[55] Note also that before the falling mass touches upon the endpoint O, its movements do not depend on its internal structure. However, its further behavior does! Try to understand the difference between the phase trajectories of a tennis ball and of a piece of plastic or used chewing gum.

Having done all these sketches, let us turn to a real object and try to draw the phase portrait of the nonlinear pendulum. As before, we first depict its movements (Figure 4.9a) on the potential diagram (Figure 4.9b, compare to Figure 4.7b). Let us first recall that the relation between the angular velocity ϕ' and the angle ϕ is given by (4.3), which we rewrite once more in a slightly changed notation:

$$\frac{(\phi')^2}{\omega_0^2} + 4\sin^2(\frac{\phi}{2}) = \frac{E}{E_0} , \quad E_0 \equiv \frac{1}{2}m\omega_0^2 l^2 .$$

If the energy is zero, the pendulum is at rest and the graph of this "movement" is the axis Ot. The image of this on the potential diagram is just one point O.

It must be intuitively clear that, for small total energy, the motion is periodic. Indeed, if $E/E_0 < 4$, the angle ϕ cannot be larger than the angle ϕ_M defined by the equation

$$\frac{E}{E_0} - 4\sin^2\frac{\phi_M}{2} = 0, \quad \phi_M < \pi.$$

Then the pendulum *oscillates* between positions $-\phi_M$ and ϕ_M. This motion is periodic, but the simple sinusoidal formula now does not work. The Huygens formula is applicable only for very small amplitudes ϕ_M. Otherwise one has to use (4.2).

The graph of this oscillating motion is the curve 1 in Figure 4.9a. In the potential diagram, Figure 4.9b, it is represented by the motion along the line 1 between the points A_1 and A_1'. At these positions, the velocity and the kinetic energy of the pendulum vanish. The phase trajectory is the closed "oval" curve 1 in Figure 4.9c, which is easy to draw by expressing ϕ' in terms of ϕ using (4.3). When the amplitude is very small, the oval becomes the circle and the graph of the motion becomes sinusoidal.

When the total energy is large, our pendulum is *rotating*. Indeed, if $E/E_0 > 4$, the kinetic energy, $(\phi')^2/\omega_0^2 E_0$, does not vanish even in the uppermost position, where $\phi = \pi$ (the potential energy is maximal for $\phi = \pi$ where it is equal to $4E_0$). It is evident that the rotation is not uniform; the velocity is maximum in the lowest position and minimum in the highest position. In Figure 4.9, this motion is represented by the curves 3.

Now we come to the most important point of our discussion. We will soon see that what is most relevant for solitons is the motion with $E = 4E_0$. In this case, the conservation of energy law gives a very simple relation between the angle and the velocity

$$\frac{(\phi')^2}{\omega_0^2} = 4(1 - \sin^2\frac{\phi}{2}) = 4\cos^2\frac{\phi}{2} . \tag{4.5}$$

The simplest "motion" of this kind is when the pendulum stays in the uppermost position. Then $\phi = \pi$ or $\phi = -\pi$, the velocity is zero, and the pendulum might stay in this position for ever, if there were no external perturbations. However, any perturbation that increases its energy, for example the slightest push, will move the pendulum and it will start rotating. If we slowly lower the pendulum (thus lowering its total energy) and then leave it, it will start oscillating. Thus

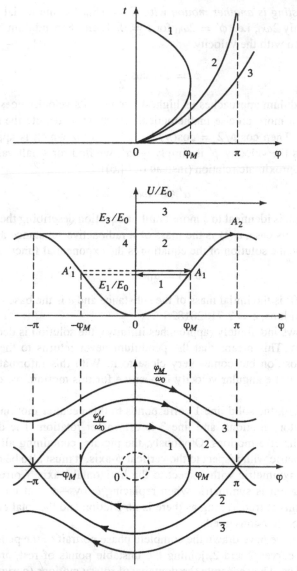

Figure 4.9 Different graphic representations of the pendulum:
a) graph of motion, b) potential diagram, c) phase portrait

the points $\phi = \pm\pi$, $\phi' = 0$ on the phase plane are unstable rest points. The motions that start arbitrarily close to these points will drive it far from them. On the contrary, the motions starting close to the stable point of rest, $\phi = 0$ and $\phi' = 0$, are oscillations — the pendulum always stays in its vicinity.

Most interesting is another motion with $E = 4E_0$. Let the initial angular velocity be exactly $2\omega_0$, i.e., $\phi' = 2\omega_0$ for $t = 0$. Then the pendulum moves to its highest position with the velocity

$$\phi' = 2\omega_0 \cos \frac{\phi}{2} . \tag{4.6}$$

When the pendulum approaches its highest position, its velocity goes to zero. To describe this in more precise mathematical terms, let us denote the small angle $\pi - \phi$ by 2α. Then $\cos \phi/2 = \cos(\pi/2 - \alpha) = \sin \alpha$ which is approximately equal to α. As the velocity ϕ' is simply $-2\alpha'$, we find for small values of α, a very simple approximate relation (instead of (4.6)):

$$\alpha' \approx -\omega_0 \alpha . \tag{4.7}$$

This equation is identical to a more familiar equation describing the radioactive decay law. In this case, $\alpha(t)$ is the mass of a radioactive substance at time t. As you may know, the solution of the equation is the exponential function,

$$\alpha(t) = \alpha_0 e^{-\omega_0 t} .$$

Here $\alpha_0 = \alpha(0)$ is the initial mass of the substance and e is the base of Napierian (natural) logarithms, $e = 2.718281828\ldots$.

When t grows indefinitely (approaches infinity), the solution is decreasing but never vanishes. This means that the pendulum never returns to the uppermost equilibrium position but comes very close to it. With this information and remembering that the angular velocity is positive for this motion, we can draw its graph.

In Figure 4.10, the solid line 1 corresponds to this peculiar motion, line 2 represents oscillatory motions, and line 3 is drawn for rotation. The dashed lines represent the inverse motions. Obviously, the picture, containing all the graphs, must be symmetric with respect to the vertical ϕ-axis. It must also be obvious that the picture is symmetric with respect to the horizontal t-axis. Correspondingly, the phase diagram is symmetric when replacing ϕ by $-\phi$. Let me stress once more that all this is true as long as there is no friction and the total energy of the pendulum is exactly conserved.

To conclude, we have drawn the complete phase portrait of the pendulum (Figure 4.9). The curves 2 and $\bar{2}$, joining the unstable points of rest, are of special importance to us. They separate the domains of rotary motions (curves 3, $\bar{3}$) from that of oscillatory (curve 1) and so are called the *separatrix* curves. We will soon see that *the shape of the Frenkel–Kontorova soliton as well as of many other solitons is described by a graph identical with the graph corresponding to the separatrix.*[56] For this reason, we shall call the solution described by this graph *the "soliton" solution* of the pendulum equation.

---

[56]The motion corresponding to the separatrix is sometimes called "asymptotic" or "limiting" motion. It is limiting in the sense that it is neither oscillatory nor rotatory, and it is transient between these two main modes.

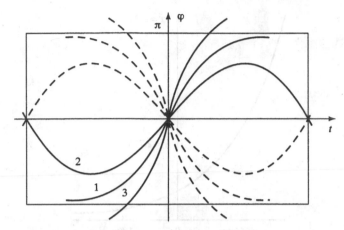

Figure 4.10 Different regimes of the pendulum motion:
1) asymptotic motion, 2) oscillation, 3) rotation

The "Soliton" Solution of the Pendulum Equation

General solutions of the nonlinear pendulum equation may be expressed in terms
of the Jacobian elliptic functions (we have mentioned that they also describe the
shape of nonlinear water waves, (Figure 2.2). Remarkably enough, to write down
the soliton solution, we need not know these higher transcendental functions. In
fact, the soliton solution described by the graph 4.10 satisfies the following elementary equation

$$\tan(\frac{\pi - \phi}{4}) = e^{-\omega_0 t} .$$ (4.8)

For this solution, $\phi(0) = 0$. The most general solution of the equation (4.6)
may be obtained by shifting the initial condition to arbitrary time t_0. In other
words, to write the general soliton solution, you may simply substitute t by $t + t_0$
in (4.8).

To better understand the behavior of the soliton solution, let us directly express
ϕ in terms of t by using the inverse to the tangent function

$$\phi = \pi - 4 \arctan(e^{-\omega_0 t}).$$ (4.9)

To draw the graph of this function (not using a PC or scientific calculator) one
may recall the graphs of the exponential function, Figure 4.11, and of the inverse
tangent function, Figure 4.12. Then it is easy to find that when t grows from $-\infty$
to $+\infty$ the solution ϕ grows from $-\pi$ to $+\pi$ and $\phi(0) = 0$.

Now recall that ϕ satisfies the differential equation (4.6). Using it and some
elementary trigonometry, one may find that

$$\phi' = \frac{4\omega_0}{e^{\omega_0 t} + e^{-\omega_0 t}} \equiv \frac{2\omega_0}{\cosh \omega_0 t} .$$ (4.10)

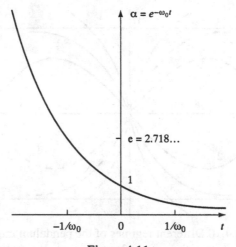

Figure 4.11

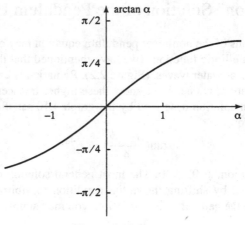

Figure 4.12

Here we have introduced the hyperbolic cosine

$$\cosh(\omega_0 t) = \frac{1}{2}(e^{\omega_0 t} + e^{-\omega_0 t}),$$

which is often encountered in soliton theory. One can easily draw the graph of this function, see Figure 4.13.

After all these preliminaries, you can draw the graphs of the functions $\phi(t)$ and $\phi'(t)$ that describe our very special motion of the pendulum, see Figure 4.14. It will be useful to memorize these two remarkable functions.

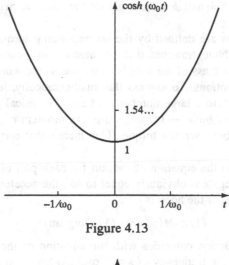

Figure 4.13

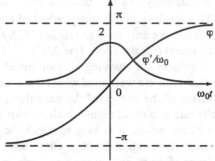

Figure 4.14 Graphs of $\phi(t)$ and $\phi'(t)$ for the asymptotic motion of the pendulum

Motions of the Pendulum and the "Tame" Soliton

One can study the main qualitative features of the pendulum motions in very simple experiments. One of the most easily accessible and good devices for these experiments is the standard bicycle wheel. To make a good pendulum, you should turn the bicycle over and fix a piece of plasticine or any other small heavy bob on the rim of the wheel. (If the wheel is not well balanced, it is better first to balance it, so that it will be at rest in any position). Now, the external force acting on the wheel is determined by the attached bob, while the body of the wheel participates in its rotations and oscillations.

Let us estimate the period of the small oscillations of this pendulum. To do this, we may imagine that the wheel is replaced by a very thin hoop, the radius of which, L, is equal to the distance from the center to the inward part of the rim and the mass is equal to the whole mass of the wheel, M. Now, the force acting on the hoop is defined solely by the bob, acting like a pendulum. It is $-mg \sin \phi$, where

m is the mass of the bob and ϕ is the angle for the effective pendulum formed by the bob.

Rotary movements are defined by the corresponding torque (moment of the forces) $-mgl \sin \phi$. Note, however, that the mass of our "wheel pendulum" does not coincide with the mass of the bob, because the whole wheel (or, now, hoop) participates in its motions. To express this mathematically, let us imagine that the hoop is divided into a large number n of small identical parts. If we apply to each part the small force $-(1/n)mg \sin \phi$, the moment of all the forces will coincide with the above written torque. This means that our "division" of the force is admissible.

Now we can write the equation of motion for each part of the hoop. As the velocity of the aggregate is obviously equal to $l\phi'$, the acceleration is $l\phi''$. Thus the Newton equation for the hoop is

$$(1/n)Ml\phi'' = -(1/n)mg \sin \phi.$$

We see that the equation coincides with the equation of the pendulum $\phi'' = -\omega_0^2 \sin \phi$, but now the frequency of small oscillations ω_0 is defined by the relation $\omega_0^2 = mg/Ml$. This result is almost independent of our approximation. In fact, to get the exact equation we only have to replace Ml by I/l, where I is the moment of inertia of the wheel (for the hoop, $I = Ml^2$).

This simple device is suitable for studying all motions of the pendulum. You only have to remember that even in a well oiled wheel there remains some friction. It causes a nonconservation of the energy of the pendulum, so that its rotations slow down and gradually turn into oscillations which, in turn, gradually die down.

By experimenting with the wheel, it is easy to check the isochronism of the small oscillations as well as its violation for large amplitudes. It is not more difficult to find the dependence of the period on the amplitude or to study qualitative features of all possible motions.

However, it is not so easy to get the experimental graphs of the motions. Of course you may make a film with a cinema or video camera, but even with these devices this is quite a job. Remarkably, there exists a much simpler experiment giving these graphs. It uses a device that, at first glance, has nothing in common with the pendulum.

To do this experiment you should find a long enough piece of thin steel wire. It must be elastic, not having much residual deformation after bending. Let us lay it on a table and try to contract it by applying a slight effort. The wire then will assume a "half sinusoidal" shape, as shown in the upper part of Figure 4.15. Let us draw tangent lines at several points and find their angles with the original straight position. The angles are counted off differently to the left and to the right of the mid-point O; the distance s of a point on the curve from O is chosen positive to the right and negative to the left of O. In this way we may "experimentally" find the dependence of ϕ on s. Theoretically it was shown that this $\phi(s)$ is an oscillatory solution of the pendulum equation (we only have to identify s with t).

Now, let us make a loop on the wire, as shown in the lower part of Figure 4.15. If the wire is long enough and is kept in the plane of the table surface, the angle

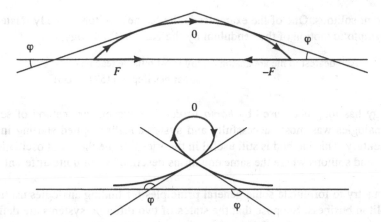

Figure 4.15 Bending shapes of a wire

ϕ grows from $-\pi$ to $+\pi$. Its dependence on s is described by the formula (4.9) after substituting s for t. The parameter ω_0 depends on the applied force F. If the wire is infinitely long and ideally elastic, the loop can freely move along the wire. This loop is one of the simplest solitons. We may call it the "tame" soliton.

Note that all bending shapes of the ideal infinitely long wire $\phi(s)$ describe motions of the pendulum. This remarkable analogy between seemingly very different phenomena was discovered[57] by the German physicist Gustav Kirchhoff (1824–1887) and is called "Kirchhoff's analogy." In fact he discovered a much more general analogy between the states of deformed elastic bodies and some motions of rigid bodies. Unfortunately, this beautiful analogy is practically forgotten. We will say more about it when we turn to Frenkel's soliton.

Concluding Remarks

A method we need for search of truth

René Descartes (1596–1650)

We are at the end of the most difficult chapter of this book. Its main thrust has been an attempt to present concepts of the general theory of oscillations, which are most important for soliton theory. We tried to illustrate these concepts with only simple but nontrivial examples. The reader who wishes to understand the structure of solitons should develop a clear understanding of linear and nonlinear oscillations of the pendulum. Especially important are the energy relations and the motion the phase trajectory of which is the separatrix (formulae (4.9), (4.19) and Figure 4.14). These solutions will allow us to understand some complex and

---

[57]The bending shapes of the wire were discovered by Leonard Euler and are called "Euler's elastics."

important solitons. One of the examples is the "tame" soliton, directly related to the asymptotic motion of the pendulum by the Kirchhoff analogy.

> And dearest to me are analogies, my most reliable teachers.
>
> Johannes Kepler (1571–1630)

Analogy has long been used by *homo sapiens*. However, the method of scientific analogies was most successfully and systematically applied starting in the 19th century. This method is still useful in theories like the theory of oscillations, waves, and solitons where the same equations describe lots of quite different systems.

Let us try to formulate some general principles for finding analogies useful in the soliton business. Suppose that the states of two different systems are defined by the same number of variables that are usually called generalized coordinates (for example, the angle ϕ for the pendulum, the charge Q for the capacitor, etc.). Suppose further that the energies of both systems, E_1, E_2, are conserved. Finally, suppose that by a transformation or renaming of coordinates and system parameters (masses, capacities, inductances, etc.) we can make the energies E_1 and E_2 identical functions of the coordinates (up to a constant multiplier). Then it is clear that the systems are completely analogous and there exists a one-to-one correspondence between their "motions" independent of the nature of the systems and meaning of the word "motion."[58]

This is a rough formulation of an analogy. Some subtleties may prove very important in using this method. For instance, the new generalized coordinates may change in different intervals. An even more important subtlety is that we may be interested in different problems for analogous systems. To be more precise, our definition of the analogy says that the systems satisfy the same equations (we have used energy only for convenience). However, as we know, to completely determine a concrete motion we have to add some auxiliary information, for example, initial values of the coordinates and velocities.

These general remarks are easier to explain by simple examples. Let us look from this viewpoint at the Kirchhoff analogy. We have mentioned that, for determining the shapes of Euler's elastics, we have to solve the pendulum equation for $\phi(s)$. Thus there is the complete analogy, and $\phi(s)$ exactly corresponds to $\phi(t)$. However, while the time variable t may change from $-\infty$ to $+\infty$, the length l of the wire is finite and so $-l/2 \leq s \leq l/2$. Now, our problem with elastics is to find the shapes of the wire for different values of the applied external force F. So, from the "pendulum viewpoint," our problem is rather odd: to find all possible movements of the pendulum starting at "time" $t = -l/2$ and ending at $t = l/2$ and to find their dependence on the parameter ω_0^2! There are other problems, which are natural for the elastics, but look very unnatural for the pendulum. For example, it is of interest to find the most stable shape of the wire for a given value of the

---

[58]We have seen that the bending shapes of the static wire are related, in this sense, to real motions of the pendulum. There exist more exotic analogies.

external force, etc. Note that, in general, the problems of the theory of elastics are more difficult to solve than natural problems for the pendulum.

An analogy need not be exact to be useful. A typical example of an approximate analogy is the relation between the standard pendulum and the cycloidal one. Using approximate analogies requires more care than exact ones. For instance, for large enough amplitudes of oscillations, the motions of the standard and cycloidal pendulums become qualitatively different. Much better, in this sense, is the analogy between the pendulum and the point mass, which is moving along the curve $y = a[1 - \cos(x/b)]$ in the gravity field that acts in the negative direction of the y-axis (this is a mathematical model of the ball rolling in a gutter). Denoting x/b by ϕ, check that small oscillations of the mass around the point $\phi = 0$ correspond to small oscillations of the pendulum having length $l = b^2/a$. Moreover, the phase portraits of these two systems are qualitatively similar. In a more precise mathematical language one may say that the portraits are *topologically equivalent.*[59] A commonly known example of such an equivalence is your image in a curved mirror.

This topological equivalence of phase portraits may be used to define a "qualitative equivalence" of dynamical systems. However, if you start to play with this, you will immediately encounter the following subtlety. All models of real systems studied by physics describe their behavior only approximately. Moreover, any mathematical model of a physical phenomenon is constructed by simplifying (idealizing) the real system. In other words, the approximation is not so much quantitative as qualitative. The more complex the system, the more serious are simplifications and more rough approximations. As Ya.I. Frenkel once said:

> ... theoretical physicist... is like a caricaturist who need not portray his model in detail... instead he must simplify, schematize it so as to expose and to stress the most characteristic features. One may (and should) demand photographic accuracy only of a theoretical description of the simplest systems. A good theory of any complex system must be only a good "caricature" of the system, which exaggerates its typical properties while intentionally ignoring other, less important ones... A good caricature of a person will not be made much better by a more accurate depiction of non-distinctive details of face and figure.[60]

Of course, most difficult is to catch these most distinctive features!

---

[59]Topologically equivalent portraits are easy to obtain by drawing one on a sheaf of thin rubber. The images obtained by stretching it without breaking are topologically equivalent. Under this transformation, closed curves remain closed, nonintersecting curves remain nonintersecting, etc.

[60]To fully appreciate this passage, one has to know that Frenkel's main hobby, from his childhood to his last days, was painting. Many portraits of his friends and acquaintances reproduce their most characteristic features, while not being caricatures. For this reason, I would prefer to compare good theories not with caricatures but with drawings of Pushkin or Picasso, who were able to express their impression of the model with just several strokes.

When we have a good understanding of the problem, we are able to clear
it of all auxiliary notions and to reduce it to simplest elements.

René Descartes

The first process... in the effectual study of sciences must be one of
simplification and reduction of the results of previous investigations
to a form in which the mind can grasp them.

James Clerk Maxwell (1831–1878)

Recognizing and neglecting immaterial details is a necessary condition for finding
a physical law. Galileo was probably the first who clearly understood this deep
truth and communicated it to others. A beautiful example of "caricaturing" is his
discovery of the law of inertia which is usually called "Newton's first law" of
mechanics (Newton attributed first and "Newton's second laws" to Galileo). The
first part of the law of inertia was known to the great Greek philosopher Aristotle
(384–322 B.C.); he clearly stated that to make a body move one should apply
some force.

It was much more difficult to imagine that the uniform motion along a straight
line does not require applying a force. Aristotle and his followers never suspected
even the possibility of such a strange law. Of course, they never could imagine
that uniform motion along a straight line might be equivalent to the state of rest,
which is the essence of Galileo's relativity principle. All experiments seem to be
against the law of inertia and relativity, and one couldn't prove them logically. To
discover this fundamental law, Galileo had to imagine a world without friction
and analyze its relation to our real world. This discovery is one of the greatest
achievements of the human mind. I don't wish to compare it with the cartoonist's
job, I would prefer to compare Galileo's vision with a creative poetic impulse,
although Galileo's work is deeper. "Wipe off its casual traits – and you will see:
the world is splendid" (Aleksandr Blok).

An approximation to the ideal inertial movement may be seen in skating (I don't
know whether Galileo was familiar with this pleasure). The main force acting on
the skater is that of gravity and it is completely balanced (neutralized) by the
vertical reaction of the ice. Now, to get the picture of ideal inertial motion one
has to imagine that the friction forces are not only small but exactly zero. Galileo
succeeded in making this last step. This is amazing, because, to the best of my
knowledge, he never studied friction experimentally.

A clear concept of friction and, more generally, of resistive media was first
formulated in Newton's *Principia*. He introduced friction (resistance) forces proportional to velocity, to the square root of velocity as well as velocity independent
forces. In Section 6 of the second volume of his famous book, he proves several theorems about pendulum oscillations in resistive media. A very beautiful
example is Theorem 21 stating that oscillations of the cycloidal pendulum with
the friction force proportional to velocity are isochronous (try to prove this theorem, it is not difficult!). Using his theory, Newton tried to find the dependence of
friction forces on velocity in experiments with a pendulum.

It is of interest to follow Newton's reasoning, and I strongly recommend that the reader look into Sections VI and VII of the second part of his book. His style is not easy for the modern reader, and to follow him in detail is almost impossible. However, the logical structure of his reasoning and his general approach look quite modern. In fact, in modern physics courses the concept of friction is introduced essentially by using a simplified version of Newton's approach. His experiments were repeated in 1915 at the suggestion of A.N. Krylov in St. Petersburg. The main design was that used by Newton, but the methods of measuring time intervals and amplitudes were improved. Newton's results were essentially confirmed. It also became quite clear that by these methods he could not establish precise enough quantitative dependencies. It was not until almost 100 years later that the final quantitative laws were discovered in the fundamental experiments of Charles Augustin Coulomb (1736–1806).

What have we learned from all these stories? The most important lesson is that an idealization of a physical phenomena is required for finding a law of physics. Once the law is established, one can explain small deviations from it by introducing "auxiliary forces," like friction. The story of discovery of the "first Newton law" and of the friction forces are typical examples of this approach. In general, finding a proper idealization may be the most difficult part of a physicist's work, leading to a deeper level of understanding. As one poet said, "Find out where's Light — you'll know where's Dark." (Aleksandr Blok)

Following Galileo and Newton, we completely ignored friction in discussing general laws of nonlinear oscillations. Is it justified in our case? At first glance, it appears that our portrait of the pendulum is only a bad caricature of the real thing. Indeed, what will happen to the portrait in Figure 4.9 if we take into account even a very small friction force? The result is shown in Figure 4.16. You immediately see that this picture is topologically not equivalent to the ideal portrait (it is enough to realize that the closed curves corresponding to oscillations become now the spirals reeling in the rest point O). So what? Is friction a characteristic feature of oscillations or not? Let us not hurry to answer.

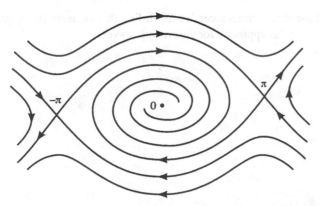

Figure 4.16 Phase portrait of a pendulum with friction

After some reflection, we may find that drawing the portrait of the "ideal pendulum" was not a quite useless pursuit. First, when the friction is small, the real phase trajectories are very close to the ideal ones. Second, and most important, is the application of the ideal pendulum to soliton theory. We will soon see that, in soliton theory, most important is the motion corresponding to the separatrix of the ideal pendulum. Thus, from the soliton point of view, friction really is an immaterial feature of pendulum motions, and adding this feature would catastrophically destroy this most important property of the pendulum.

And now the final remark. For describing pendulum motions we have used some mathematics. Was it necessary? Well, if we were concerned with the real pendulum, we would survive with energy conservation, simple geometric considerations, and a qualitative picture extracted from simple observations. However, this equipment is not adequate for treating asymptotic motions, and the mathematics introduced above is really necessary. To what extent it is necessary is a different question. In this connection, let me quote a very relevant passage from Mandelshtam's *Lectures on Oscillations*. After presenting, in a very transparent manner "without any mathematics," oscillations of a massive point in a trough he reflected:

> In such a simple picture everything follows from intuitive considerations. Then, why we have made several mathematical deductions in the previous lecture? The reason was that all "simple minded" talks based on "obvious" facts are wrong in at least one point. Let the kinetic energy of the bob be less than the potential one. We know that in this case the bob stops at some height. But are we sure that this will happen in a finite time interval? And then, what is the answer to this question for the asymptotic motion?... Here the visual arguments are useless and a mathematical investigation is necessary; without it you will not find any serious answer.... To know to what extent you must be mathematically rigorous is the most difficult problem for the physicist. We may better say that he must correctly determine what is the measure of the mathematical rigorousness adequate to the problem he is studying.

I can only hope that I have succeeded in finding this measure in my presentation and will try a similar approach for describing waves.

Chapter 5

From Pendulum to Waves and Solitons

> The sea waves have a melody
> And there is harmony in Nature...
>
> Fyodor Tyutchev

Everyone has an intuitive understanding of waves on water. However, from the scientific viewpoint, these waves are extremely complex objects. For this reason we will first study less known but much simpler waves. In this, we follow Newton who, in his study of the waves of sound, introduced a basic model of waves in chains of particles bound by springs. Although these waves are only caricatures of the real waves in water, studying them will allow us to understand basic properties of all waves.

Newton's idea was that in the wave, travelling through a medium, all particles of the medium oscillate like small pendulums that are coupled to neighboring pendulums. To make a mathematical description possible, he further simplified some hypotheses. In a modernized form, these hypotheses may be formulated as follows. Suppose that the deformations of the moving small particles of the medium may be neglected and that we may regard them as mass points. Suppose further that the coupling of the neighboring particles may be represented by massless springs. This idealized system is, in fact, a very good model of simple crystals. However, waves in all media have many features in common and, in this sense, the model is applicable to all waves. As this model is still too complicated for a first acquaintance, we further simplify it by reducing to one space dimension.

Waves in Chains of Bound Particles

Let us consider a chain made of identical point particles of mass m that are bound by ideally elastic, identical, and massless springs. Allowing these particles to move only along a straight line, we get the model of a *one-dimensional crystal*. So let us call the point masses "atoms." The quotation marks should remind you that these "atoms" have yet no relation to the real atoms (in what follows we often omit the quotation marks).

Let the length of each spring be a. Then, labelling the atoms by integer numbers n, the coordinates of the equilibrium positions of the atoms will be $x_n^0 = na$, see Figure 5.1. The coordinate of the nth moving atom is $x_n(t)$, or simply x_n. Denoting the deviation of the nth atom from its equilibrium position by $y_n = x_n - x_n^0$, we may visualize the deviation by points with the coordinates (na, y_n) — look at the top of Figure 5.1. Here the dependence of the deviation on the coordinates is represented by a smooth curve $y(x)$ passing through the points (na, y_n). In the configuration of the atoms, shown in our drawing, we may approximate the smooth curve just by connecting the points with rectilinear segments. However, in general, this piecewise linear curve will have a zigzag shape. It must be obvious that such configurations are not related to the waves in a continuous medium. Thus we will consider only those configurations in which deviations of all neighboring atoms are so correlated that the piecewise linear curve is smooth enough, meaning that the differences of the slopes of the neighboring linear segments are very small. A more precise geometrical criterion is that the dependence of the angles α_n (defining these slopes) on n is weak enough. Note that the angles themselves need not be small, only the *differences between the successive angles must be much less than the angles.*

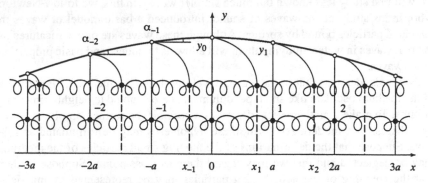

Figure 5.1 Chain of particles bound by springs:
Below: the chain in the equilibrium state.
Above: a deviation from the equilibrium state and its graphical represenation
by a smooth curve $y(x)$

The smooth curves $y(x)$ of our model may correctly represent the deviations of small particles of "one-dimensional" continuous media (like air in the organ tubes, the violin string, etc.), if the number of the "atoms" in the chain is very large and the distances between them are very small. In addition, the characteristic size of the wave (the wavelength for periodic waves) must be large in comparison to a, to meet the just mentioned conditions of smoothness. Later in this chapter, the relation between discrete chains of particles and corresponding continuous objects will be discussed in more detail. Now let us turn to a qualitative description of general properties of waves.

To observe waves on your table, you may construct a very simple device. For this you need a long enough rubber band and large paper clips, or something similar. Then look at my very schematic image of the device in Figure 5.2 and make it! Of course, this chain is far from ideal and, in fact, it is much more complex than the chain of atoms. The main shortcoming of our "clips device" is a substantial dissipation of the energy of the waves due to friction forces inside the rubber. On the other hand, its important virtue is the rather small velocity of the wave propagation. In addition, you may change this velocity by making the clips heavier — it should be intuitively clear that the speed is smaller for heavier clips.

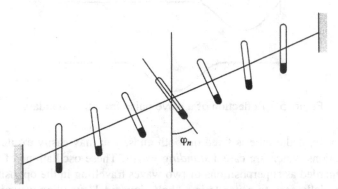

Figure 5.2 Clips on a rubber band – a simple device for observing waves

If you fix each clip at its center of mass, your system will be equivalent to the linear chain of atoms.[61] The angle ϕ_n plays the role of y_n and, instead of the masses of the "atoms," we have to use the clips momenta of inertia. The torque created by the twisted rubber band plays the role of linear tension of the springs. In all, the analogy here is the same as between linear and torsional small oscillations.

Our device has one new feature as compared to the infinite chain of atoms — the presence of boundaries, in which the rubber band is fastened. This results in reflection of waves from the boundaries, as shown in Figure 5.3. This picture represents several successive positions of the atoms in the chain having fixed boundaries. To further visualize the reflection process, use a long rubber tube. In this case, Figure 5.3 represents not just graphs but the transverse waves themselves.

---

[61] Otherwise the force of gravity creates an additional momentum and each clip becomes a nonlinear pendulum coupled to its neighbors. We will use this device in the next chapter.

The experiments with the waves in a rubber tube are not difficult to perform and the reader may exercise her/his fantasy and inventiveness in studying them.

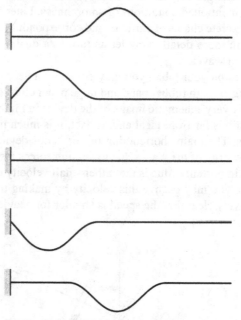

Figure 5.3 Reflection of a wave pulse from the boundary

By the way, if the tube is fixed on both ends, you may easily excite stationary oscillations which are called *standing waves*. These oscillations of the tube may be regarded as superpositions of two waves travelling in the opposite directions. Especially easy to generate is a "half-sinusoidal" standing wave in which all the points of the tube have the same phase and the amplitude is maximal at the midpoint. By definition, the length of this standing wave is equal to the doubled length of the tube.[62] Slightly more difficult to produce is the standing wave in which the midpoint of the tube is not moving. Then you will see the whole sinusoidal vibration and, naturally, the length of this wave is equal to the length of the tube. Somewhat easier to generate is this wave in our chain of clips. You just make the chain oscillate in any mode and then suddenly fix the central clip. It is not necessary to give you further instructions on experiments with waves; so we will continue discussing their general properties.

The standing waves in bounded systems are also called the *normal modes* of vibrations of the system, or *modes*, for short. In one-dimensional systems with fixed ends (like our simple models, or guitar strings) we can enumerate them by the number of half-waves N between the ends. Modes with small N are called

---

[62]Below we will show that the length of the standing wave is equal to the length of the travelling waves of which it is "constructed."

lower modes, the ones with large values of N are, naturally, *higher* modes. The lowest mode having $N = 1$ is called the *main* or *principal* one. The main mode is easiest to excite and longest to live. Higher modes die faster because of higher energy losses due to friction. For this reason they are more difficult to observe.

It is most convenient to start studying modes using the chain of atoms with fixed ends. One atom constrained by two strings is just the old linear oscillator studied in the previous chapter. So the simplest chain consists of two atoms and three springs, see Figure 5.4. The equilibrium positions of the atoms are $x_1^0 = a$, $x_2^0 = 2a$, and the springs are fixed at $x_0 = 0$, $x_3 = 3a$. Let us write and solve the equations describing motions of the atoms.

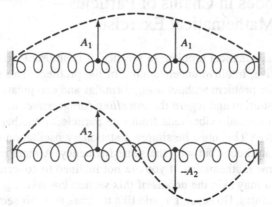

Figure 5.4 Chain with two particles and modes of its oscillations

First we introduce a new notation. In the preceding chapters we used only the derivatives in time, which were denoted by the prime superscripts. Now we are dealing with systems extended in space (usually one-dimensional) and so we need a notation for the derivative in the coordinate at fixed time (space-derivative). For this reason, from now on we will denote the time-derivative by the "dot," reserving the "prime" for the space-derivatives to be introduced later. Thus, the velocity of the nth mass point in the chain is $\dot{y}_n$ and its acceleration is $\ddot{y}_n$.

It is easy to write Newton's law for the chain consisting of two particles:

$$m\ddot{y}_1 = -ky_1 + k(y_2 - y_1), \quad m\ddot{y}_2 = -k(y_2 - y_1) - ky_2. \qquad (5.1)$$

Indeed, the first spring acts on the first mass with a force proportional to its dilation y_1, the factor k depending on the elasticity of the spring.[63] The minus sign says that, when the spring is stretched ($y_1 > 0$), this force is directed to the left. The extension of the second spring is obviously $y_2 - y_1$ (the negative sign means that the spring is contracted) and so the force is $k(y_2 - y_1)$. In the same way, you may write the force acting on the second mass.

---

[63] We suppose that the springs are ideally elastic and satisfy Hooke's law.

It is not difficult to solve these equations by simply adding and subtracting them. Then we get for $z_1 = y_1 + y_2$ and $z_2 = y_1 - y_2$ the familiar equations

$$m\ddot{z}_1 = -kz_1, \quad m\ddot{z}_2 = -3kz_2,$$

the general solution of which is known from Chapter 4. It is a nice exercise for you to find the dependence of y_1 and y_2 on time from the general solution. In the next section we will solve this problem by a more complex method, which can be applied to determining motion in the chains with any number of particles.

Finding Modes in Chains of Particles – Important Mathematical Exercise

To solve any mathematical problem in physics, it is advisable to start by trying to get a qualitative, intuitive understanding of the problem. This means solving at least part of the problem without using formulas and computations. Let us follow Newton's intuition and regard the *travelling (progressive) wave* as a process of propagating harmonic vibrations from one particle to another. What is then the standing wave? This must be simply a stationary oscillation of all particles with different amplitudes but with the same frequency. A more precise description requires some mathematics. If you are not inclined to concentrate on this at the moment, you may skip the details of this section but take a good look at the pictures and formulas. However, I would like to stress that this section deals with one of the fundamental principles of physics and not physics alone.

Suppose now that all particles in the chain oscillate harmonically with the same frequency ω, and let us try to see what we can make of this. We know that for any harmonic motion with the frequency ω, the acceleration is proportional to the distance from the equilibrium point, in our case, this means that $\ddot{y}_n = -\omega^2 y_n$. Then we may substitute these expressions for the accelerations in the equations of motion, thus excluding the accelerations. In the two-particle case, the equations are just (5.1), and so we get the following simple system of two linear equations for two variables y_1 and y_2:

$$(\omega^2 - 2\omega_0^2)y_1 + \omega_0^2 y_2 = 0, \quad \omega_0^2 y_1 + (\omega^2 - 2\omega_0^2)y_2 = 0. \tag{5.2}$$

Here $\omega_0^2 \equiv k/m$ and ω is the frequency of the oscillation we wish to find.

It is clear that the system (5.2) has the trivial solution $y_1 = y_2 = 0$. Forget about it and concentrate on a solution for which either y_1 or y_2 are not identically zero. Let us thus suppose that $y_1 \neq 0$, use the first of the equations to express y_2 in terms of y_1, and substitute the resulting expression into the second equation. We find that

$$[(\omega^2 - 2\omega_0^2)^2 - \omega_0^4]y_1 = 0. \tag{5.3}$$

As $y_1 \neq 0$, the expression in the square brackets must vanish. This gives the condition

$$(\omega^2 - 2\omega_0^2) = \pm\omega_0^2,$$

and thus ω may assume one of the two following values:

$$\omega_1 = \omega_0, \quad \omega_2 = \sqrt{3}\omega_0. \tag{5.4}$$

These frequencies are called the *fundamental frequencies* of the system, (or *eigenfrequencies*).

Now, for $\omega = \omega_1$, (5.2) says that $y_2 = y_1$. Similarly, for $\omega = \omega_2$ we have $y_2 = -y_1$. If you recall that y_1 and y_2 obey the equations $\ddot{y}_n = -\omega_n^2 y_n$, you immediately find possible solutions $y_1(t), y_2(t)$. For $\omega = \omega_1 = \omega_0$

$$y_1 = y_2 = A_1 \cos[\omega_1(t - t_1)], \tag{5.5a}$$

and for $\omega = \omega_2 = \sqrt{3}\omega_0$

$$y_1 = -y_2 = A_2 \cos[\omega_2(t - t_2)]. \tag{5.5b}$$

Here, the constants A_1 and A_2 are arbitrary amplitudes. The time moments t_1 and t_2 are also arbitrary; they define the phase of the oscillation.

The obtained solutions represent two possible modes of oscillations in our simple system. In Figure 5.4, they are shown schematically. The dashed lines represent the corresponding modes of the transverse vibrations in the rubber tube. Of course, this correspondence looks somewhat superficial, but what should you expect from a sketch drawn with just two points!

We may use the linearity of our main equation (5.1) and write the general solution as the sum of the solutions (5.5a) and (5.5b)

$$y_1 = A_1 \cos[\omega_1(t - t_1)] + A_2 \cos[\omega_2(t - t_2)], \tag{5.6a}$$

$$y_2 = A_1 \cos[\omega_1(t - t_1)] - A_2 \cos[\omega_2(t - t_2)], \tag{5.6b}$$

This linear combination of the two modes can be reduced to simple harmonic modes only if $A_1 = 0$ or $A_2 = 0$. Otherwise, for arbitrary amplitudes and phases, it is not even periodic.

Exercise. Consider the motions of both masses when $A_1 = A_2 = 1$ and $t_1 = t_2 = 0$. Show that the periodicity condition $y_1(T) = y_1(0)$ and $y_2(T) = y_2(0)$ cannot be satisfied for any T. Will the motion, described by (5.6), be periodic if, say, $\omega_1 = 1$ but $\omega_2 = 3$?

It is obvious that formulas (5.6) represent the most general possible oscillation of our system because, by properly choosing the amplitudes A_1, A_2 and the phases given by t_1, t_2, one can make the initial coordinates $y_1(0)$, $y_2(0)$ and velocities $\dot{y}_1(0)$, $\dot{y}_2(0)$ arbitrary numbers. In fact, the unknown constants A_n and t_n can always be expressed in terms of the initial positions and velocities. But we need not know these more complicated expressions.

The most important and remarkable thing is that the general motion of the masses in our model is represented by a linear combination of simple harmonic modes. In this sense, the motions of two coupled oscillators are reduced to the motions of two completely independent ones! Even more remarkable and important is that this statement is also true for any number of particles in the chain! This can be shown by a fairly simple generalization of the above arguments.

As an exercise, try to find the fundamental frequencies of the chain made of three particles. The result is

$$\omega_1^2 = \omega_0^2(2 - \sqrt{2}), \quad \omega_2^2 = 2\omega_0^2, \quad \omega_3^2 = \omega_0^2(2 + \sqrt{2}).$$

The modes themselves are schematically presented in Figure 5.5. The exact meaning of this picture is that the M-th mode ($M = 1, 2, 3$) can be written in the form

$$y_n^{(M)} = A_M \sin\left(\frac{\pi}{4} M n\right) \cos[\omega_M (t - t_M)], \quad n = 1, 2, 3.$$

For any fixed M, this expression represents the motion of the nth mass in the chain. Notice that the modes in the two-particle chain may be rewritten in the same form

$$y_n^{(M)} = A_M \sin\left(\frac{\pi}{3} M n\right) \cos[\omega_M (t - t_M)],$$

where $M = 1, 2$ and $n = 1, 2$.

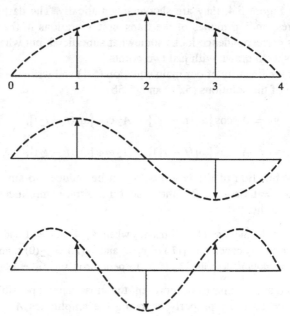

Figure 5.5 Modes in the chain with three particles

Now you may guess the general pattern and write for the Mth mode in the N-particle chain a similar expression

$$y_n^{(M)} = A_M \sin\left(\frac{\pi M n}{N + 1}\right) \cos[\omega_M (t - t_M)], \tag{5.7}$$

where $M = 1, \ldots, N$ and $n = 1, \ldots, N$. As above, A_M and t_M are arbitrary constants that should be defined by the initial conditions. To find the fundamental frequencies ω_M you have to substitute our hypothetical solution into Newton's equation for the particles in the chain. It is easy to write them by generalizing the two-particle equation (5.1). For the nth atom, the equation is

$$m\ddot{y}_n = k(y_{n+1} - y_n) - k(y_n - y_{n-1}), \qquad (5.8)$$

where, as above, $k/m = \omega_0^2$. This equation is also valid for the atoms at the ends of the chain, if we define $y_0 = y_{N+1} = 0$.

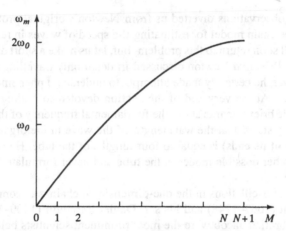

Figure 5.6 Continuous interpolation of the formula (5.9) for the spectrum of the frequencies of free oscillations in the N-particle chain

Now, to find the fundamental frequencies, we substitute (5.7) into (5.8) and replace $\ddot{y}_n^{(M)}$ by $\omega_M^2 y_n^{(M)}$. Then, to derive the expression for ω_M, you have to use some standard identities for the trigonometric functions. The computations are not very difficult but tedious, and I write only the final result

$$\omega_M = 2\omega_0 \sin\left(\frac{\pi M}{2(N + 1)}\right). \qquad (5.9)$$

This formula defines the *spectrum* of the frequencies for all free (i.e., unforced) oscillations (modes) in the chain. Naturally, the number of possible frequencies is equal to the number of the particles. The dependence of the frequency on M is visualized in Figure 5.6 above. For the low-frequency modes (with small M), the dependence is practically linear. The high-frequency modes depend on M non-linearly, and the highest possible frequency is $2\omega_0$. The origin of this "frequency limit" is intuitively clear: if one of the atoms is forced to oscillate very fast, the neighboring atoms can't follow its motion and the wave can't propagate. You can check this in simple experiments.

Let us stop at this point. Frankly, I am afraid that few readers will be sufficiently patient to follow the arguments of this section in detail. For advanced readers, it will appear too elementary, and for beginners, too difficult. But, for those who wish to understand nonlinear phenomena, a good knowledge of linear science must be acquired. The expansion of arbitrary motion into normal modes is surely one of the most fundamental principles of linear physics. You will see it in action many times.

A Historic Digression. Bernoulli Family and Waves

These simple observations diverted us from Newton's original problem. In fact, he invented the chain model for estimating the speed of waves in real continuous media. We will soon return to this problem, but, let us make a short digression into history. In his *Principia* Newton discussed in detail only travelling (progressive) waves. However, he certainly made attempts to understand phenomena related to standing waves. At the very end of the section devoted to finding the speed of sound, he made brief remarks about the fundamental frequency of the organ tube. Then he simply stated that the wavelength of the wave in the organ tube (which is open at one of its ends) is equal to four lengths of the tube. He obviously was not aware of other possible modes in the tube and never formulated the idea of a standing wave.

The theory of oscillations in the one-dimensional chain was completed by Johann Bernoulli (1667–1748) and his son Daniel Bernoulli (1700–1782). Daniel and Johann's brother Jacob were the most prominent scientists belonging to the famous Bernoulli dynasty of scientists. In the 16th century, the family emigrated from Antwerp, which had been conquered and terrorized by Spain. Eventually, the Bernoulli family settled in Switzerland, in the quiet and beautiful city of Basel. Jacob and Johann were Leibniz's students and became distinguished mathematicians. Daniel, with his friend Leonhard Euler, studied mathematics under guidance of his father.

The Bernoulli family and Euler had close relations to Russia. From 1725 to 1733, Daniel was working in St. Petersburg. Euler came to him in 1726, and he spent a good half of his life there. At that time, the Russian czars generously supported science, and the Academy of St. Petersburg soon became one of the renowned European centers of science. D. Bernoulli and Euler were elected members of the Academy and published in its Proceedings many of their scientific discoveries.

Normal modes were discovered by Johann and Daniel Bernoulli. The representation of arbitrary motions in the chain by linear combinations of normal modes — *the principle of superposition of oscillations and waves* — was established by Daniel Bernoulli. He was the best physicist among the Bernoullis. Most celebrated are his discoveries in hydrodynamics, in the kinetic theory of gases and in the theory of oscillations. By the way, the relations between the members of the Bernouilli family were not always benevolent. From time to time, rather harsh disagreements between them became known to the public. Especially unpleasant were discussions of priority. What matters is the fact that historical studies of science reveal a deep truth — each step strongly depends on many preceding ones, even though scientists do not always appreciate the role of their predecessors.

The superposition principle, as natural as it seems, met serious objections of other scientists. Even Euler and Lagrange were among its opponents. They themselves came very near to discovering the superposition principle but were afraid of serious mathematical problems that would arise if it were valid. These problems were resolved much later. We will soon say more about them.

The one-dimensional chain served as a model for study by Lagrange and Cauchy of waves of sound in solids, liquids, and gases. Also a major contribution to the theory of chains of atoms was made by Kelvin at the end of the 19th century. He developed a theory of waves in chains consisting of several kinds of atoms, not just identical ones. Then he applied his theory to propagating waves of light in solids and to explaining the phenomenon of *light dispersion*.[64] The dispersion of light was discovered in the 17th century by the Czech scientist Jan Markus Marzi. Newton rediscovered it in his famous experiments with glass prisms on the spectral resolution of light. Dispersion is so important for soliton theory that we will study it in some detail later in this chapter. Just now, let us return to the end of the 18th century and try to follow the most important developments in the theory of waves after Bernoulli.

D'Alembert's Waves and Debates about them

> Imagination is no less important for a geometrician than for a poet in his moments of inspiration.
>
> Jean D'Alembert (1717–1783)

Euler started his investigation of oscillations in strings without using a discrete approximation. He regarded the string as a one-dimensional continuous medium and used the equations appropriate for continuous media. In this approach, the motions of the string are represented by a function $y(t, x)$, the values of which give the deviations of each point of the string (having the coordinate x) from its equilibrium position ($y = 0$) at time t. The equation, describing motions of the string, depends not only on derivatives in time t but also on derivatives in the space variable x. Such equations are called *partial differential equations*, because the derivatives of functions of two variables on each of the variables, with the other one kept fixed, are called *partial derivatives*. Euler was the first to start

---

[64]In optics, the word dispersion usually refers to the phenomena caused by the dependence of the refraction index on the frequency or wavelength of the light wave. The most beautiful example of dispersion phenomena is a rainbow. In the general theory of waves, dispersion usually refers to dependence of the velocity of a propagating wave on its length (we will see that the optical concept of dispersion is not different), while the relation between frequency and wavelength is called the dispersion formula.

a systematic study of partial differential equations, describing strings. He also introduced and studied some partial differential equations of hydrodynamics, but these are extremely difficult to deal with because they are essentially nonlinear.

Unlike motions of water, those of strings obey a very simple linear partial differential equation. Using Euler's ideas, the French mathematician and one of the creators of the first Encyclopedia[65] Jean Le Rond D'Alembert (1717–1783), obtained in 1748 its general solution

$$y(t, x) = f(x - vt) + g(x + vt). \tag{5.10}$$

This solution is called the D'Alembert solution, or D'Alembert's wave. Here, f and g are completely arbitrary functions of one argument and thus D'Alembert's wave may reproduce any motion in the ideal string[66] in terms of the right-moving wave $f(x - vt)$ and the left-moving wave $g(x + vt)$ (left and right movers).

Let us choose, for example, $g = 0$. Then, the solution is a wave propagating to the right with velocity v. Note that the velocity depends on the elasticity and tension of the string and is the same for all waves in a given string. If we now specify the dependence of f on its argument x as $f(x) = \sin(2\pi x/\lambda)$, we get the sinusoidal travelling wave

$$y(t, x) = \sin\left[\frac{2\pi}{\lambda}(x - vt)\right] = \sin[\kappa(x - vt)] = \sin(\kappa x - \omega t).$$

Here we have introduced the *wave number* $\kappa \equiv 2\pi/\lambda$ and the familiar *circular frequency* $\omega \equiv 2\pi\nu = 2\pi v/\lambda$. The last equality is the well-known relation between the frequency and the wavelength of the travelling wave. Note that $T \equiv 1/\nu$ is the period of the oscillations in each point of the wave while λ is its period in space.

Obviously, the D'Alembert wave can describe arbitrary wave pulses, like the one shown in Figure 5.3. It is also easy to get standing waves in the string. For instance, with $f(x) = g(x) = \sin(\kappa x)$, we have $y(t, x) = 2\sin(\kappa x)\cos(\omega t)$, which is a harmonic standing wave. In general, one can reconstruct D'Alembert's left and right movers, if the initial positions and velocities of all points of the string are known (this means that y and $\dot{y}$ are given functions of x at $t = 0$). In this sense, the arbitrary functions f and g replace the arbitrary parameters A_M and t_M of the discrete chain. We will also call the latter a "discrete string," to underline the similarity between strings and chains.

This similarity is not quite obvious. In fact, Euler and Lagrange regarded continuous and discrete strings as very different objects. The reason was that, accord-

---

[65]With Denis Diderot he was the soul and director of the collective work on a monumental *Encyclopedia of Sciences, Arts and Handcrafts)*. 33 volumes of this "Bible" for the future irreligious times were published from 1751 to 1777.

[66]As this solution describes *any* wave, we may temporally forget about the equation. Its role was just to select, among functions of two variables, a class of functions representing waves. Of course, in due time, we shall recall D'Alembert's equation. The main equations, describing solitons, are closely related to it.

ïng to the principle of superposition, the general (arbitrary) motion of the string may be represented by a sum (superposition) of harmonic modes. From this, one can infer that an arbitrary function of one variable can be represented as a sum of trigonometric functions. Euler and Lagrange thought this highly improbable and, accordingly, adhered to the opinion that the superposition principle is valid for discrete systems, consisting of a finite number of particles, but is not applicable to continuous systems like strings.

The apparent paradoxes were finally resolved by Fourier in 1807. Fourier proved that any function, defined on a finite interval, can be represented as an infinite sum of trigonometric functions similar to the modes described above. This generalization of the mode expansion is called the *Fourier expansion* or the *Fourier series*. It is interesting that Fourier's work was strongly stimulated by Euler's and Lagrange's investigations. I think that Lagrange's repudiation of the superposition principle is really somewhat strange and needs a psychological explanation. In fact, it was he who first clearly established the relation between oscillations of discrete and continuous strings!

Now let us try to write the D'Alembert equation. Otherwise it will seem like we are dealing with a mysterious character in a play about whom everyone is talking but noone has really seen. You might even think that the equation is very difficult to write and even more difficult to comprehend. In fact, it is very simple. To obtain it, let us first visualize its solution (5.10) by drawing graphs of two arbitrary functions $f(x)$ and $g(x)$ of one variable x on separate sheets of transparent paper, and then moving them with constant speed in opposite directions along the x axis. Let me call this procedure a "cinematic" representation of the D'Alembert waves.

This "cinematics" (please don't mix it with kinematics) is easy to translate into mathematical language and thus write the equation for the waves. Recalling the definition of the derivative we may write

$$\dot{f}(x - vt) \approx [f(x - v(t + \Delta t)) - f(x - vt)]/\Delta t,$$

$$f'(x - vt) \approx [f((x + \Delta x) - vt) - f(x - vt)]/\Delta x.$$

Here Δx and Δt are arbitrary infinitesimally small numbers (i.e., small with respect to any scale of our physical problem). Let us choose $\Delta x = -v\Delta t$. Then we find the relation

$$\dot{f} + vf' = 0,$$

which is independent of Δx and Δt and is the differential equation for the right-moving wave. Quite similarly, you may obtain the equation for the right mover $g(x + vt)$. It is $\dot{g} - vg' = 0$. These are nice equations, but we need one describing both waves simultaneously.

To derive this equation, you may observe that the x-derivative and the t-derivative of f also depend only on $x - vt$. Thus, they also obey the equation for left movers, which means that f satisfies two equations:

$$\ddot{f} + v\dot{f}' = 0, \quad \dot{f}' + vf'' = 0.$$

By excluding from these equations $\dot{f}'$, we find that $\ddot{f} = v^2 f''$. Applying the same reasoning to g, you may prove that it satisfies the same equation. As the differentiation operation is linear, the sum $y = f + g$ also obeys it. Thus the *D'Alembert wave equation* is

$$\ddot{y} + v^2 y'' = 0. \tag{5.11}$$

Of course, we have only shown that any sum of the left-and right-moving waves (5.10) satisfy D'Alembert's equation. In fact, it would be not very difficult to prove the inverse statement and thus demonstrate that equations (5.10) and (5.11) are indeed equivalent. Let me leave this as an exercise to mathematically oriented readers.

This simple equation and its generalizations to two and three space dimensions play a fundamental role in the physics of continuous media. Its applications are difficult even to list. It is applicable not only to the waves in piano or violin strings but, practically, to all linear waves. As Russell would say — from waves in "oceans of water, air and ether" to waves describing elementary particles.

If you think about all this, it is really amazing that the transition from the equation describing oscillations of one particle (in our new notation it is $\ddot{\phi} = -\omega_0^2 \phi$) to the continuous system, "consisting" of a "continuum" of particles, proves to be so simple both conceptually and technically. Physicists have become so accustomed to this equation that nobody is astonished by its effectiveness. However, if you try to mentally grasp the whole world of natural phenomena behind this simple formula, the insight it gives, and the innumerable human inventions based on it, then you will agree that its effectiveness is really fantastic. Mathematics is a grand beauty, but its efficiency in dealing with nature is really difficult to comprehend.

On Discrete and Continuous

> Between separate existing things there are other things
> and between them, in turn, there exist more. And thus Entity is infinite
> Zeno from Elea (490–430BC)

Let us return to equation (5.8), which was born not in the land of Muses but in the grim environment of "point masses on springs." Yet, we know that piano strings and our poor masses, tied in one chain, are close enough relatives. In fact, with a large enough number of masses and springs, the chain may approximate the string as precisely as you wish. It was Lagrange who first established this fact in 1754. However, many problems remained unsolved before they were clarified by Cauchy in 1830.

In his exhaustive work, Cauchy explicitly demonstrated how to find the evolution of the string from any initial state, which is defined by initial positions and velocities of all points of the string (mathematicians call this the "Cauchy problem"). He also found the relation between the D'Alembert and Fourier solutions of the wave equation; then he proved the validity of the superposition principle. This was the final stroke of mastery on the problem. Especially beautiful and important was his study of the dispersion of waves. He demonstrated very clearly and explicitly that, for waves that are much longer than the distance between the atoms, the velocity of waves does not depend on the wavelength and thus there is no dispersion in this case. For short waves, the velocity may significantly depend on the wavelength, and Cauchy obtained the mathematical expression for this dependence (the Cauchy dispersion formula). The Cauchy theory is fully applicable to waves in elastic media. He also tried to apply it to waves of light, but there the theory gave only a qualitative description of the dispersion. This was not a failure of his mathematical theory but rather a deficiency of the physical model. As we have mentioned above, a better model for waves in crystal-like structures was proposed much later by Kelvin.

Why is it so important to know the precise relation between discrete and continuous strings? Or, more generally, between the Newtonian discrete medium and the Eulerian continuous medium (by the way, the word "discrete" is of Latin origin, *discretus* means discontinuous, separated). The main reason, for us, at the moment, is very simple — some problems are easier in the discrete picture, others — in the continuous one. Thus, studying waves of sound in a crystal, one may often forget about its atomic structure and use the continuous approach. Yet, there may exist effects of atomic structure even for long waves. For example, the properties of a real three-dimensional crystal may be different for waves travelling in different directions. It is also easy to understand that the atomic structure is even more important for waves of light.

Well, it is better not to try to embrace everything, so let us concentrate on simple things in our simple "body-spring" model. To be more specific, suppose that we wish to use it for an approximate treatment of longitudinal waves in an elastic rod (in the same way one can treat waves in organ tubes, piano strings, etc). The problem is to find how discrete equations, like (5.8), may be related to continuous wave equations, like the D'Alembert equation. The general idea of the transition to the continuum limit is fairly clear. First, we have to make point masses lighter and the springs shorter, but keep constant the average linear mass density (the average mass distributed over the interval of unit length, $\rho_1 = m/a$). Second, we also have to keep constant the elasticity of the springs.

Let us define the second condition in more precise terms. To do this, we need a clearer and more precise understanding of the elasticity of the springs and the rod. In the right-hand side of (5.8), we have written the expression for the force, acting on the nth mass from the nth spring, in the form $F = k(y_{n+1} - y_n) = k\Delta l$. Here, the spring parameter k depends not only on its elasticity but also on its length. Now, when the spring's length is increased by Δl, both its halves become longer by $\Delta l/2$. As the force is the same along the string, this means that, for the spring of length $a/2$, the spring parameter is $2k$ instead of k. The message of this simple observation is clear: we have to rewrite the force $k\Delta l$ as $ka\Delta l/a$. Then the new elasticity parameter ka will be independent of the length of the spring. In other words, for a rod of any length, the same relation between the applied force and the resulting relative increase of its length is valid, namely, $F = K(\Delta l/l)$. If we choose $ka = K$, our "body-spring" model will reproduce the elastic properties of the rod.

Let us find the velocity of waves in the chain (and thus in the rod it approximates). Suppose that a wave of a fixed shape (or a wave pulse) is moving along the chain with the speed v. This means that it runs over the distance a for the time $\Delta T = a/v$. Then, looking at Figure 5.7 showing the position of the wave at three successive moments, you easily find that

$$y_{n-1}(t) = y_n(t + \Delta t), \quad y_{n+1}(t) = y_n(t - \Delta t).$$

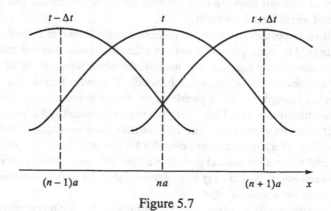

Figure 5.7

Let us, for a moment, assume that $y_n(t)$ is the coordinate of a moving point. Then it has the velocity $\dot{y}_n(t)$ and acceleration $\ddot{y}_n(t)$. In the continuum limit, the distance a and the time interval Δt must be chosen small; more precisely, a must be much smaller than the wavelength, [67] i.e., $\Delta t \ll \lambda/v$, where λ is the wavelength (or a characteristic size in case of the wave pulse). Now for small enough time intervals, the acceleration of the moving point is approximately constant and we may apply Galileo's law

$$y_n(t \pm \Delta t) \approx y_n(t) \pm \dot{y}_n(t)\Delta t + \ddot{y}_n(t)\frac{(\Delta t)^2}{2}. \tag{5.12}$$

Expressing now y_{n-1} and y_{n+1} in terms of $\dot{y}_n$ and $\ddot{y}_n$ and inserting the result into the righ-thand side of (5.8), you will find the relation $[m - k(\Delta t)^2]\ddot{y}_n(t) = 0$. This means that $(\Delta t)^2 = m/k$ and thus the velocity of the wave is

$$v = a\sqrt{k/m} = \sqrt{(ka)/(m/a)} = \sqrt{K/\rho_1}.$$

---

[67]In this respect, our Figure 5.7 is somewhat misleading. To make the picture comprehensible, we have drawn a wave that actually does not satisfy these conditions.

These simple arguments may be used to derive the D'Alembert equation directly from (5.8). We first write $y_n(t) = y(t, x = na)$. Then, it is easy to understand that the approximation (5.12) is also applicable to $y(t, x \pm \Delta x)$, where $x = na$ and $\Delta x = a$ (in the right-hand side of (5.12) $y'(t, x)$ and $y''(t, x)$ will now replace $\dot{y}_n(t)$ and $\ddot{y}_n(t)$). Substituting these approximate expressions into (5.8), it is not difficult to find the approximate equation for $y(t, x = na)$:

$$\rho_1 \ddot{y} = K y''. \tag{5.13}$$

In the continuum limit, when a vanishes but K and ρ_1 are kept fixed, this becomes the exact D'Alembert equation for the function $y(t, x)$. In accord with our previous result, the velocity of the D'Alembert wave is $v = \sqrt{K/\rho_1}$.

At first glance, our formula for the speed of sound depends not only on the material of the rod but also on its cross section S. Indeed, our linear density of the rod is the product of the standard volume density ρ on the cross section, $\rho_1 = \rho S$. However, recall that K is numerically equal to the force required to make the rod two times longer: when $\Delta l = l$ we have $F = K$. The force is obviously proportional to the cross section, and thus we may write $K = ES$. The new parameter E depends only on the elastic properties of the material; it is called the Young modulus (introduced by Thomas Young) or modulus of elasticity. Thus the final formula for the speed of sound is

$$v = \sqrt{E/\rho}. \tag{5.14}$$

Modules of elasticity and volume densities for different materials are well known; you can find them in physics and engineering reference books. For instance, steel has $\rho = 7.8 \text{g/cm}^3$, $E = 2.1 \cdot 10^{12} \text{g/(cm} \cdot \text{s}^2)$. Thus the speed of sound in steel is $v = \sqrt{E/\rho} \approx 5$ kilometers per second.

Could you suggest a simple measurement of this velocity? It is much easier to measure not the velocity itself but a related quantity, in this case, the wavelength $\lambda = v/\nu$. For waves in steel, generated by a vibrator with frequency $\nu = 10\text{kHz}$, we thus have to measure the wavelength $\lambda \approx 50\text{cm}$.

The Speed of Sound and How it was Measured

Before the end of the 18th century, scientists generally believed that the transmission of sound through solid bodies was instantaneous. The first measurement of the speed of sound in solids was attempted by a German experimentalist Ernst Chladni (1756–1827). As a matter of fact, he measured the ratio of the velocities of sound in solids and in gases. He also made the first precise measurements of the speed of sound in different gases. For this, he used organ tubes (think why!).

Chladni graduated from a law faculty, but was more interested in natural sciences which he studied by himself. Influenced by the works of Bernoulli and Euler, he was particularly impressed with acoustics and decided to confront the theory of acoustic phenomena with reality. He started his experiments with "sounding plates" and discovered the "sound figures" ("sound pictures").[68] Chladni's pictures may be seen on thin vibrating plates. Have you seen them? No? Why not try!

Chladni discovered and studied many phenomena related to acoustic vibrations and invented several musical instruments on which he played himself. His experiments were always ingenious and clear; you can find modernized versions of them in textbooks. He had written the first systematic monograph on acoustics, which was published in 1802. Undoubtedly, Chladni was a brilliant lecturer and made good advertising of his experimental acoustics. Napoleon Bonaparte was so impressed by Chladni's lectures and demonstrations that he granted 6000 Francs for translating *Acoustics* into French.

Let us return to waves of sound. The speed of sound is easy to estimate by using dimensional considerations. Recalling what we know about the waves of sound, it is not difficult to figure out that the speed of sound in a rod may depend on the Young modulus E, the rod volume density ρ and, possibly, on the wavelength λ. Thus we may try to write the formula for the speed of sound, $v = dE^a \rho^b \lambda^c$, where a, b, c, d are some unknown constants. Using the dimension formulas $[E] = ML^{-1}T^{-2}, [\rho] = ML^{-2}, [\lambda] = L$, and $v = LT^{-1}$, you can easily find that $a = -b = 1/2, \ c = 0$. Thus, $v = d\sqrt{E/\rho}$, where d is a number that can't be derived solely from dimensions (above, we have found that $d = 1$, see (5.14)).

As you see, the dimensional considerations show that the speed of sound cannot depend on any power of the wavelength λ. This means that dimensional analysis is insufficient for deriving dispersion (which is, by definition, a dependence of the speed on the wavelength). The same remark is applicable to a possible dependence of v on the amplitude of the wave. Fortunately, this dependence is negligibly small for small amplitudes, when the waves are approximately linear, and in this chapter we never leave the safe ground of linearity.

By the way, when there is no dispersion, dimensional analysis tells you that the speed of the sound waves does not depend on their amplitude. To check this, take $c = 0$ in the above dimension formula for v, but introduce a dependence on the amplitude.

Similar dimensional considerations allow us to estimate the speed of sound in liquids, say in water. In this case, the Young modulus plays the role of volume elasticity K. (I hope you will not confuse this with the K of the preceding section). It is defined by the relation $\Delta p = -K \Delta V/V$, where Δp is the increment in pressure required for compressing the volume V of the liquid by ΔV. This formula essentially coincides with the Young formula for the rod, if we rewrite it as $F/S = E(\Delta l/l)$. Using this analogy, we immediately find the speed of sound in liquids: $v = \sqrt{K/\rho}$. For water, $\rho = 1\text{g/cm}^3, \ K \approx 2.13 \cdot 10^{10}\text{g/cm} \cdot \text{s}^2$

---

[68]It was Galileo who first made the waves of sound visible. He used a wine glass in water (with its top above the surface) and observed radial surface waves produced by its vibrations. I have read about this but failed to reproduce visible waves. Galileo was a genius indeed (and probably used much better wine glasses)! Robert Hooke was the first to observe and systematically study sound figures but they became widely known after Chladni's popular demonstrations.

and thus $v \approx 1460 m/s$. Note that the speed depends on the density, which changes with temperature.

Until the end of the 18th century, it was believed that sound waves did not propagate through liquids. Chladni was of the opinion that they do propagate, but didn't try to substantiate it by experiments. The first measurement of the speed of sound in water was made in the year of Chladni's death. Swiss scientists Jean Colladone and Jean Sturm (1802–1892) measured it in Lake Geneva (Lake Leman). Their result was $v = 1435 m/s$, for temperature $8°C$.

Using the above arguments, you may easily estimate the speed of transverse waves in stretched strings. In this case, the force in Newton's equations for the particles of the string is proportional to the string tension force F. If the curvature and the dilation of the string are small, its elasticity is irrelevant. This means that one may try the formula $v = dF^a \rho_1^b \lambda^c$, where ρ_1 is the linear density of the string. Equating the dimensions of both sides, we find $v = d\sqrt{F/\rho_1}$. From the dynamical considerations (or making experiments), you could find that $d = 1$. This result is applicable to strings of any nature – to piano and violin or guitar strings, to strings made of cotton or nylon (e.g., fishing lines).

The nylon string is most convenient for experiments. Changing the tension of the string, you may change the frequency of the lowest mode, which defines the main tone. Using a piano or a tuning fork, you may identify it with a definite musical note. Then you will know the frequency of sound produced by the string. For that you have to know that the "la" of the first octave is normally tuned by the tuning fork to the frequency $v_0 = 440 Hz$. The frequencies v_n of other notes may be found from the formula $\log(v_n/v_0) = (n/12)\log 2$. For instance, for the "la" in the second octave $n = 12$, and thus the frequency is $2v_0$. To the left of the first octave, you have to use negative n.

So, with the aid of a piano or any other well tuned musical instrument, you can find the main frequency of a string and then determine the velocity of waves by using the relation $v = \lambda v$ (remember that λ is the doubled length of the string). This very simple idea allowed Chladni to measure the speed of sound in many solids and gases. Instead of a piano he used a "monochord" that was probably invented by Pythagoras. "Chord" is simply the Greek word for string and, in fact, monochord is just one string with fixed ends. It is attached to a resonator and is supplied with a simple device permitting its length to change.

Now you can easily derive, by analogy, a formula for the speed of sound in gases. From the Boyle–Mariotte relation between the pressure and the volume, $pV = $ const, we derive a relation similar to that used above for liquids, $\Delta p = -p(\Delta V/V)$ (mass and temperature are fixed). Thus, in gases, the volume elasticity, K, plays the role of pressure p. This gives the Newton formula for the speed of sound in gases $v = \sqrt{p/\rho}$. To derive this formula, Newton used much more intricate arguments based on his discrete model for waves. It seems that they became more or less comprehensible only much later, after the work on waves by Bernoulli, Euler and Lagrange. They certainly knew Newton's result and eventually found a better theory of waves in continuous media outlined above. Yet the waves of sound in gases remained a stumbling block to the theory of waves for many years.

The main deficiency of the Newton formula for the speed of sound in gases is that it disagrees with measurements.[69] Newton knew that his formula gives an incorrect value of the speed of sound in air, but his attempts to explain the discrepancy were vague and unconvincing. The experiments of Chladni made the difficulty even more apparent. The discrepancy persisted for all gases. Let us estimate it for air. Write the Boyle–Mariotte relation in the standard form $p/\rho = RT$ where R is the standard gaseous constant, and T is the temperature in the Kelvin absolute scale. Taking $T = 273°K = 0°C$, we get $v \approx 280$m/s, instead of the well-known value $v \approx 330$m/s.

The origin of this discrepancy, as well as the correct theory, was finally found by Laplace. His arguments were remarkable and basically simple. You know well that air warms up if you compress it fast enough (say, when pumping the tire of your car); conversely, it cools off when decompressed quickly. However, the Boyle–Mariotte law gives the relation between pressure and density when the temperature is kept fixed. Thus it is natural to suppose that the Boyle–Mariotte law is not applicable to sound waves when compressing and decompressing are done very quickly. Laplace assumed that the compressed and decompressed regions in the wave have no time to exchange heat energy and suggested using the adiabatic law instead of the Boyle–Mariotte law. From this assumption he derived the correct formula. Although his theory was disputed for about 30 years, this did not undermine the firm status of the theory of waves. In the beginning of the 19th century its basic laws were established and its applications expanded rapidly to new fields.

This was especially important for the wave theory of light. Fresnel's wave optics succeeded in explaining almost all known phenomena and predicted new ones to look for in experiments. The only failure of the wave theory was its apparent inability to explain *dispersion of light*. Dependence of the speed of light on the material in which the waves propagate was naturally explained (as in the theory of sound waves). However, nobody succeeded in explaining the well-established dependence of the speed of light waves on their wavelength in the same medium. Poisson, even after the famous experiment of Arago, mentioned in Chapter 1, was not convinced and remained doubtful about wave optics. His main objection concerned exactly the dispersion problem. In his reply to Poisson's objections, Fresnel pointed to the molecular structure as a possible source of optical dispersion. But he did not elaborate his ideas and soon died. It was Cauchy who developed and completed the Fresnel remark on dispersion.

---

[69] The first precise measurement of the speed of sound in the air was accomplished in the collective work of several members of the Academy of Sciences of Paris. They measured the time for the sound of gun fire to travel through the distance of 30km. To diminish the unavoidable influence of winds they simultaneously fired two guns separated by that distance.

Dispersion in Chains of Atoms

The relation of dispersion to atomic structure is best understood in our chain model. Though it describes waves of sound rather than waves of light, the nature of the phenomenon is the same. This was the starting point of Cauchy's approach too. Let us follow him and find a dispersion formula for the waves in our chain of "atoms." Recalling what we have learned of the relation between waves in the discrete chain and the elastic rod, let us attempt to write the travelling wave solution of all equations (5.8) ($\omega = 2\pi v$, $\kappa = 2\pi/\lambda$, $v = v\lambda$):

$$y_n = A \sin(\omega t - \kappa na) = A \sin[\kappa(vt - na)].$$

If you now substitute, as we did above, na by x and y_n by $y(t, x)$, you will get the familiar harmonic travelling wave. Its velocity $v = \Delta x/\Delta t = \omega/\kappa$ may be found from the condition for its phase ($\omega t - \kappa x$) to be constant. For this reason this velocity is called the *phase velocity*. If you run with the phase velocity of the wave, you will see no motion in it. Now, for our wave $\ddot{y}_n = -\omega^2 y_n$ and the equation (5.8) relates y_n to the neighboring y-s:

$$-\frac{\omega^2}{\omega_0^2} y_n = (y_{n+1} + y_{n-1}) - 2y_n . \tag{5.15}$$

Using the trigonometric formula expressing the sum of two sines as a product of two other sines), you easily find for our sinusoidal wave

$$y_{n+1} + y_{n-1} = 2y_n \cos(\kappa a) .$$

Inserting this into equation (5.15), we see that it will be valid for all n if and only if

$$\omega^2 = 2\omega_0^2 [1 - \cos(\kappa a)] = 4\omega_0^2 \sin^2(\pi a/\lambda) . \tag{5.16}$$

This is the *Cauchy dispersion formula*.

For a wavelength much less than the interatomic distance, that is, when $(\pi a/\lambda) \ll 1$, we may approximate $\sin(\pi a/\lambda) \approx \pi a/\lambda$ and get $\omega \approx 2\pi \omega_0(a/\lambda)$. This means that the dispersion in this limit disappears as the velocity becomes independent of the wavelength. Indeed,

$$v(\lambda) = \omega\lambda/2\pi \approx a\omega_0 = a\sqrt{k/m} = v.$$

This result is already known to us. We see once more that in the continuum limit, when $a \to 0$, the dispersion vanishes. The dependence of the phase velocity on the wavelength becomes apparent when the interatomic distance is not too small in comparison with the wavelength. Using the definition of the phase velocity and the Cauchy relation, we obtain

$$v(\lambda) = \frac{\omega\lambda}{2\pi} = v\frac{\sin(\pi a/\lambda)}{(\pi a/\lambda)} . \tag{5.17}$$

This important relation shows that the velocity decreases with decreasing λ.

Note that this is meaningless to apply to very short waves, when we are interested in the continuum limit. To describe the real dispersion phenomena in continuous media, we then may approximate the exact Cauchy formula (5.17) assuming that the interatomic distance a is smaller than the wavelength λ but not negligibly small (say, $a < 0.1\lambda$). Then, a good enough approximation for the sine function in the Cauchy formula is $\sin\alpha \approx \alpha - \alpha^2/6$, and inserting this in (5.17), we obtain

$$v(\lambda) \approx v\left(1 - \frac{\pi^2}{6}\frac{a^2}{\lambda^2}\right).$$

This approximation is sufficient for understanding the essence of dispersion and it gives a qualitative explanation of the dispersion of light waves in material media. However, the ratio of the distances between real atoms to the wavelength of light is too small to explain the observed dependence of the refraction index on the wavelength. It follows that the Cauchy dispersion formulae can't give the precise, quantitative description of the optical dispersion phenomena.[70]

In addition, there exists "anomalous" dispersion which was discovered in 1862 but completely explained only in the 20th century. It is related to internal excitations in molecules and thus has to be treated in terms of Quantum Mechanics. The full quantitative treatment of the normal dispersion is also based on Quantum Mechanics.

The Cauchy approach is in general not applicable to surface waves on water. We soon shall study them in some detail but first let me return for a moment to standing waves and introduce the Fourier expansion. It generalizes the Bernoulli expansion in the modes to continuous media.

On How to "Perceive" the Fourier Expansion

> The grand piano was open wide
> And strings in it were trembling...
>
> Fyodor Tyutchev

Recalling the Bernoulli treatment of modes in the discrete string, you may observe that the Cauchy formula is the continuous limit of our expression (5.9) for the frequencies of the modes. To see this, recall that the wavelength of the mode number M is $\lambda_M = 2(N + 1)a/M = 2l/M$. Inserting this in (5.9) you obtain the Cauchy relation between the frequency ω_M and the wavelength λ_M of the mode.

---

[70]The dispersion is studied by measuring the dependence on λ of the refraction index $n(\lambda) = c/v(\lambda)$ where c is the speed of light in a vacuum. The wavelength of red light is two times larger than that of violet, while its refraction index in transparent glass is 1% less. Thus, according to the Cauchy formula, a/λ must be approximately 0.1 but in reality the ratio of the interatomic distance to the wavelength is less than 10^{-3}.

It is now clear how to write the modes in a continuous string or an elastic rod. We simply replace na by x in our expression (5.7) for the mode in the chain and denote $y_n(t)$ by $y(t, x)$. Thus the mode number M in the rod is

$$y_M(t, x) = A_M \sin(\kappa_M x) \cos[\omega_M(t - t_M)]. \qquad (5.18)$$

Here $\kappa_M \equiv \frac{2\pi}{\lambda_M}$ and the frequency of the mode is given by the Cauchy relation (5.16) with λ_M replacing λ. In the continuous limit we have to express ω_0 in terms of the phase velocity, $\omega_0 = v/a$. Then, recalling that for small $\pi a/l$, $\sin(\pi a/\lambda)$ is $\pi a/\lambda$, you immediately find that $\omega_M = 2\pi v/\lambda_M = v\kappa_M$. Thus, for an elastic rod, the speed of sound v is given by (5.14) and the frequency of the Mth mode is[71]

$$\nu_M = \frac{\omega_M}{2\pi} = \frac{M}{2l} \sqrt{\frac{E}{\rho}}. \qquad (5.19)$$

You can similarly find the frequencies of modes in piano strings, organ tubes, etc.

Of course, all modes satisfy the D'Alembert wave equation (if you are not sure, please check this now!). Any linear combination of the modes also is a solution. This approach to solving the D'Alembert equation was known to Bernoulli. What he did not know, and what bothered Euler and Lagrange, was that in this way he could find all solutions of the D'Alembert equation. This fundamental fact was established by Fourier, who finally proved that the approaches of D'Alembert and Bernoulli give the same results. One has only to take into account that the number of modes in the continuous case is infinite! That was most important and was the main mathematical difficulty because "infinity" is not an obvious matter.

The representation of any vibration of the string as a sum of an infinite number of modes and other similar expansions (for instance, an expansion of the travelling wave in a sum of sinusoidal travelling waves, harmonics) are naturally and justifiably called Fourier expansions. It is symbolically written as

$$y(t, x) = \sum_M y_M(t, x). \qquad (5.20)$$

For an elastic rod or piano string with fixed ends, explicit expressions for the modes are given by the formula (5.18), where the arbitrary amplitudes A_M and phases ωt_M are to be defined by initial conditions (as in the Bernoulli expansion). The *Fourier Theorem* states that by adjusting the amplitudes and phases you may expand any vibration.

Usually the amplitudes decrease rapidly for increasing M. For instance, the main mode of the piano string is identified by our ear as giving the tone (say, "la" of the first octave). Other modes (overtones) that also are excited define the timbre. We do not hear very high modes for two reasons. First, because their amplitudes are small and, second, because our ear is not sensitive to frequencies

---

[71]I hope the reader is not confused by my carelessly calling the circular frequency simply "frequency."

above 20kHz (for this reason, higher tones are poor in timbre). Thus very often we may completely forget about higher modes and use just several of the most important lower modes in the Fourier expansion.

Now it is not difficult to imagine what physical processes allow us to listen to Beethoven's last piano sonata #32. The pianist excites the modes of the strings which, in turn, generate travelling waves, harmonics.[72] The Fourier expansion of the travelling wave in harmonics keeps all information on the Fourier expansion of the string. As the waves of sound are linear, and there is no significant dispersion, this information encoded in the amplitudes and phases of the harmonics reaches our ears essentially undistorted. Then the resonators[73] in our ears respond (resonate) to corresponding harmonics. To be more precise, each harmonics excites proper oscillations in the resonators so that the spectrum of the frequencies, the amplitudes and the phases of the beautiful Beethoven accords are reproduced. Then nonlinear processes start to develop in our nervous system, in our brain, in our muscles. Finally our soul resonates to Beethoven's soul, to the universal primordial spirit of his music ... But this has nothing to do with physics ...

Even the simplest linear part of the described process is more complex. You know that the piano sounds differently in a small room vs a concert hall. By the way Beethoven himself could not normally listen to his late compositions. Sometimes he used a walking stick that transmitted the sound waves directly from the piano resonating deck to his mighty skull, and then to internal resonators in his ear. Complications aside, we may say that our capability to perceive music essentially depends on the fact that sound waves are linear and do not suffer dispersion. Having linearity, we can apply the Fourier expansion (in this case we expand any wave into harmonics). Having no dispersion, all the harmonics run with the same velocity. As a result, the wave moves with that velocity without any distortion. More generally, in any medium small amplitude waves travel undistorted if, in addition, there is no dispersion. Sound waves are so long that their dispersion is practically negligible. For waves of light this is not true, which is one of the reasons why the mechanism of vision must be much more complex than that of hearing.

Dispersion of Waves on Water Surface

> The water is an example to us, an example ...
>
> Wilhelm Müller (1794–1827)

---

[72]Sometimes, sinusoidal modes are also called harmonics, which is especially natural when you talk about music. I prefer to use harmonics to mean only travelling sinusoidal waves.

[73]The origin of this term is in the Latin verb "sonare" which means to sound. Of the same root is sonata. Our world is full of resonators that respond to the external harmonics having frequencies equal to their proper frequencies. So the piano strings may tremble or even "resonare"— you know what I mean.

Dispersion is easy to observe for waves on the water surface. In deep water the velocity of waves is proportional to the square root of their lengths. This is Theorem 37 from the third book of Newton's *Principia*. Then Newton derives a formula for their velocity. He compares vertical oscillations of the particles of water to oscillations of a pendulum having length $l = \lambda/2$. For one full oscillation (of period T), the wave translates by the distance λ, and thus $v = \lambda/T = \sqrt{g\lambda/\pi}$ (using Galileo's formula). Taking into account circular motions of the particles of water, one could derive the correct result $v = \sqrt{g\lambda/2\pi}$. Of course, this simple derivation is applicable only to waves on deep water, when λ is much less than the depth h. In the opposite limit, when λ is much larger than h, the velocity depends only on h, and there is no dispersion. The exact formula in this case is $v = \sqrt{gh}$.

These results, up to numerical factors, may be obtained by dimensional considerations and simple physical ideas. Obviously, the velocity depends on λ, h, and g. Simple observations show that there is no significant dependence of the velocity on the amplitude of the wave. It is also easy to understand that the velocity is independent of the density of water. The gravity force acting on a particle of water is proportional to its mass and thus disappears from Newton's equation (if it seems not to be convincing, you may keep ρ in the dimension formula).

With all this, we may write $v = dg^a\lambda^b h^c$ and find a, b, c by comparing the dimensions. Thus we get

$$v = d\sqrt{g\lambda}(h/\lambda)^c.$$

Here the dimensionless constants d and c remain undetermined, and we have to use our grey cells, which tell us that probably only particles that are close enough to the surface are really moving. This means that for large enough depth, there is no dependence on it and thus, in this limit $c = 0$. Some clever cells may suggest that in the opposite limit, when $h \ll \lambda$, there must be no dependence on the wavelength. In fact, the movements of particles in the wave become similar to those of atoms in our chain or in the rod. Of course, this statement is somewhat vague, but it suffices to conclude that for shallow water we must take $c = 1/2$.

The exact theory gives a formula valid for all h and λ. It says that the velocity grows monotonically with the wavelength but has the upper limit $v = \sqrt{gh}$. The approximate expressions for short and long waves are:[74]

$$v = \sqrt{g\lambda/2\pi} \text{ for } \lambda < 2\pi h; \tag{5.21a}$$

$$v = \sqrt{gh}\left(1 - \frac{(2\pi h)^2}{6\lambda^2}\right) \text{ for } \lambda > 2\pi h. \tag{5.21b}$$

The first formula is a good approximation for wavelengths somewhat less than $2\pi h$. For larger λ, the second formula becomes applicable. You may see that the corresponding dispersion is analogous to the Cauchy dispersion and depth plays the role of the interatomic distance.

---

[74]The exact formula is $v^2 = g\tanh(hk)/k$ where $k \equiv 2\pi\lambda$ and the hyperbolic tangent is defined in the Appendix.

It is worth noting that the term "shallow water" must not be understood literally, in an everyday sense. For very long waves generated by earthquakes in the ocean its average depth (about 5 km) is small enough. These waves, known in Japanese as "tsunami," are in fact typical, and very dangerous, solitons. We will make acquaintance with them in the next part of this book. Here we only mention that the range of sizes and velocities of these waves is enormous. Waves with wavelength much larger than 5km may generate terrible damage when coming to the shore. They have velocities around $\sqrt{gh} = 800$ km/h. In a cuvette for development of photographs, waves much longer than the average depth of 0.5 cm have velocity around 20 cm/s. It is not difficult to generate a "micro-tsunami" in the cuvette by knocking it, and the velocity is very easy to measure. You may also create at your table "deep waters" and make many observations of surface waves and experiments with them. I leave this to your initiative and inventiveness.

When checking the "deep-water" formula, you might find that it gives incorrect results for short waves having wavelength less than about 5cm. The reason for the discrepancy is the influence of surface tension. To understand its role, let us first neglect gravity. Then the surface tension defines the total restoring force acting on the particles of water. To estimate the velocity of the waves in this approximation, let us use dimensional analysis. The surface tension T is measured by the energy required to increase the surface by a unit. For pure water $T \approx 0.072$ Joule/m^2. The velocity may also depend on the density ρ and the wavelength λ (we suppose that the amplitude is so small and the depth is so large that dependence on them may be neglected). Using the standard approach, you may find that $v^2 = dT/\rho\lambda$. Here the dimensionless number d cannot be found from dimensional considerations. The exact theory first proposed by Kelvin (1871) gives $d = 2\pi$. These waves are called *capillary waves* (recall that surface tension is responsible for the rise of water in capillaries). As their velocity is inversely proportional to the wavelength, their dispersion law is anomalous, in optical usage.

Kelvin had found a very interesting effect of surface tension, namely that there exists a minimal velocity for deep water waves. The reason is not difficult to understand. Indeed, by inspecting all the above expressions for the velocity of waves, you may observe that the square of the velocity is proportional to the restoring force (represented by the elasticity k for springs, the tension F for strings, etc). Then it is natural to suppose that in the case of the two forces — gravity and surface tension — we may just add them and write

$$v^2 = \frac{g\lambda}{2\pi} + \frac{2\pi T}{\rho\lambda}.$$

In fact this is the correct Kelvin formula. The dependence of v on λ is presented in Figure 5.8. It is easy to understand that the velocity is minimal when the two forces are equal. By equating both terms in the above formula, we find $\lambda_{min} = 2\pi\sqrt{T/(\rho g)}$. For pure water $\lambda_{min} \approx 17$ mm and the minimum value of the velocity

$$v_{min} = v(\lambda_{min}) = (4Tg/\rho)^{1/4} \approx 23 \text{ cm/s}.$$

The expression for v^2 is convenient to rewrite in the form

$$v^2 = \frac{1}{2} v_{min}^2 \left(\frac{\lambda_{min}}{\lambda} + \frac{\lambda}{\lambda_{min}} \right). \tag{5.22}$$

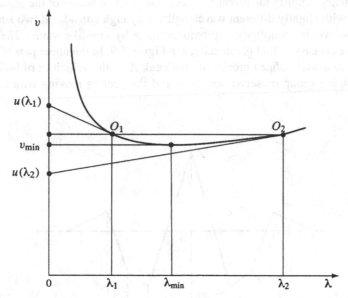

Figure 5.8 Dispersion law for capillary waves— the dependence of the phase velocity v on the wavelength λ

Capillary waves are familiar to everybody. You may produce them in a basin by letting droplets of water fall from a pipette. You will observe that the higher above the surface the pipette is placed, the longer are the waves. To see that the short waves are indeed capillary, add some soap to the water in the basin. Then the surface tension decreases, making the short waves slower, while the velocity of the long waves remains unchanged.

On the Speed of a Pack of Waves

We never observe infinite sinusoidal waves but only finite-size groups of waves. Russell, probably the first to systematically study these groups, found that the velocity of the whole group is in general less than the velocity of the peaks moving inside it. Apparently, the waves inside the group move through it and disappear at the front of the pack. This phenomenon was explained in 1876 by Stokes, who introduced the concept of *group velocity*.[75] A year later, Rayleigh undertook a

---

[75] For waves in discrete lattices, this concept was actually invented by Hamilton, who reported his ideas to the Irish Academy of Sciences in 1839 and later published two short communications. In fact

more general approach to the problem. He found how the group velocity is defined by dispersion and obtained a beautiful expression for group velocity.

In deriving Rayleigh's formula we follow his book *Theory of Sound*, which is the first and one of the best books on the general theory of vibrations.[76] First, let us maximally simplify the problem by considering two waves of the same amplitude but with slightly different wavelengths. Rayleigh considered two sinusoidal waves, but we, for simplicity, approximate them by saw-like waves. The sum of such waves is easy to find graphically; see Figure 5.9. In the upper part of the figure, we see a well-defined group with the peak A. If the velocities of both waves are equal, the group preserves its shape and the peak is moving with the same velocity.

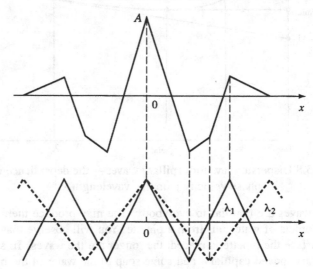

Figure 5.9 Wave packet (above) formed by superimposing
two saw-shaped waves (below)

Now suppose that there is a dispersion, in which case the velocities of the two waves are different. Let, for definiteness, $v_1 = v(\lambda_1) < v_2 = v(\lambda_2)$. What shall we see in this case? Let us draw the graphs of motion of our two waves, Figure 5.10. It must be clear that at the moment t_0, the original shape of the group is reproduced, but the peak is shifted to the point x_0. Thus it is natural to call $u = x_0/t_0$ the group velocity.

---

he had developed a fairly complete theory of the group velocity, but this became known only 100 years later, after his archive had been published.

[76]Rayleigh's contemporaries failed to realize that with this book, a general theory of waves and oscillations was born. Even Helmholtz regarded it simply as a very good book on acoustics.

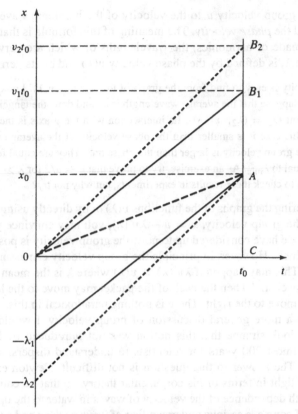

Figure 5.10 Graphical representation of the uniform motion of the crests
of the two waves in the packet of Figure 5.9. The straight line OA
represents the uniform motion of the crest of the wave packet

To find it in terms of the known velocities v_1, v_2 and the wavelengths λ_1, λ_2, we observe
that $AB_1 = \lambda_1$ and $AB_2 = \lambda_2$. By examining Figure 5.10 you easily find that

$$v_1 t_0 - x_0 = \lambda_1, \quad v_2 t_0 - x_0 = \lambda_2.$$

Let us define the average velocity of the waves in the packet, $v = (v_1 + v_2)/2$, and the
average wavelength, $\lambda = (\lambda_1 + \lambda_2)/2$. Define also the differences of the velocities and the
wavelengths, $\Delta v = v_2 - v_1$, $\Delta \lambda = \lambda_2 - \lambda_1$. Then by adding and subtracting the two above
relations we can find that

$$t_0 \Delta v = \Delta \lambda, \quad v t_0 - x_0 = \lambda.$$

Excluding from these equations t_0 and x_0 we obtain the relation $(v - u)/\lambda = \Delta v / \Delta \lambda$.
As the differences Δv and $\Delta \lambda$ are small we may approximate the right-hand side by the
derivative $v'(\lambda)$.

The exact Rayleigh formula may thus be written in the beautiful form

$$(v - u)/\lambda = v'(\lambda). \tag{5.23}$$

It relates the group velocity u to the velocity of the harmonic wave $v(\lambda)$ that is usually called the *phase velocity*. The meaning of this formula is that the velocity of a packet made of harmonics, the wavelengths of which are very close to an average value λ, is defined by the phase velocity $v(\lambda)$ and by its derivative $v'(\lambda)$.

Group velocity is easy to find from the graph of the function $v(\lambda)$. Look, for instance, at Figure 5.8. Suppose that the average wavelength is λ_2 and draw the tangent to the curve $v(\lambda)$ at the point $O_2 = (\lambda_2, \ v(\lambda_2))$. Its intersection with the y-axis is the group velocity $u(\lambda_2)$. In this case it is smaller than the phase velocity. If the average wavelength is $\lambda_1 < \lambda_{min}$, the group velocity is larger than the phase one. They are equal for the average wavelength equal to λ_{min}. As an exercise, try to prove that $u \approx v/2$ for $\lambda \gg \lambda_{min}$. A more difficult task is to check these results in experiments, but why not try?

By considering the graphs of the functions $v(\lambda)$ or by directly using the Rayleigh formula for the group velocity, $u = v - \lambda v'(\lambda)$, you may convince yourself that for all waves we have considered up to now, the group velocity is positive (please try to prove this!). However, in principle the group velocity can assume also negative values. This may happen if $\lambda v'(\lambda) > v(\lambda)$ where λ is the mean value of the waves in the packet.[77] Then the peak of the packet may move to the left while the waves inside move to the right. There is nothing paradoxical in this.

Turning to a more general discussion of group velocity, I would like to say that it might look strange that this notion was not introduced by Newton and that it took almost 200 years for scientists to understand dispersion and all its implications. The answer to this question is not difficult. Newton explained the dispersion of light in terms of his corpuscular theory, so that he could not relate the wavelength dependence of the velocity of waves in water to the optical dispersion. Then, of course, a serious understanding of wave packets and group velocity might arise only after developing a general theory of dispersion phenomena. Thus the notion of group velocity became widely recognized only at the end of the last century. After that, the concepts of dispersion and of group velocity were applied to many branches of physics and became standard. Nevertheless, even today some applications of group velocity may seem difficult for inexperienced students.

L.I. Mandelshtam in his final, unfinished lectures of 1944 gave a very clear and intuitive answer to possible questions about this rather difficult concept.

> The notion of velocity has its origin in the theoretical description of particle motions. It is sensible and clear, provided that the particle can be identified, i.e., if we can state that we have the same particle at different points of space. In the case of waves, we deal with translating the state rather than the particle. To define a velocity, we thus must have a way of identifying the state. In a dispersionless medium ... excitations move without changing their shapes and thus can be identified. However, when there is a dispersion, moving excitations are deformed and ... we have to define what is the velocity of a given sort of the excitation. For example, there is no unique definition of the velocity of a cloud. This might be, say, the velocity of its

---

[77]For instance, this condition is satisfied for the dispersion law $v(\lambda) = d\lambda^b$ if $b > 1$.

edge or something else . . . . A similar uncertainty exists with the velocity of the dispersing excitation.

Above we define . . . the notion of group velocity which is only meaningful for the group preserving its shape. In dispersive media this is not true. However, under a certain condition . . . the deformation of the group is slow and . . . group velocity approximately describes the movement of the group. Anyhow, in dispersive media . . . we may define different velocities, and group velocity is one of several possible definitions.

In contrast, the velocity of a free moving soliton can be defined uniquely. The soliton preserves its shape and so its velocity is defined as exactly as the velocity of the particle. A soliton is not a cloud! But then, what is the soliton from the viewpoint of dispersive wave theory? Well, neglecting nonlinearity, we may expand the soliton in harmonics. The small nonlinearity results in pumping the energy from faster to slower harmonics, and thus it may happen that this interaction between harmonics equalizes the diffusion effects of dispersion. This is what makes the Russell soliton stable.

How Much Energy is Stored in a Wave?

Before turning to our main objective — solitons — we have to answer one more question: What is the energy of waves and what happens to this energy when a wave or a group of waves is travelling in space? It is easy to answer this question using our body-spring model. The energy is obviously the sum of the kinetic energies of the masses and of the potential energies of the springs. It is not difficult to derive this energy for any given wave. For infinite periodic waves, the total energy is infinite, and we call the energy of the wave the energy stored on the wavelength. The energy of a localized wave group (or of a soliton) is usually more difficult to calculate but it is sufficient for us to know that it is finite and concentrated in a finite domain of space.

Let us derive the energy of a long wave in our body-spring system (Figure 5.1). The kinetic energy of the nth mass is $m\dot{y}_n^2/2$. A periodic wave with amplitude A and wavelength $\lambda = 2\pi/\kappa$ is

$$y_n = A\sin(\omega t - \kappa na).$$

It is easy to understand that, for this wave, the energy is independent of time. So let us calculate it for $t = 0$. As $\lambda \gg a$, we may write $\lambda = Na$ where N is a large integer number. Then the energy of the nth mass is

$$\frac{1}{2}m\omega^2 A^2\cos^2\left(2\pi\frac{n}{N}\right)$$

and the energy of the wavelength $\lambda = Na$ is the sum of the energies of the atoms with numbers from 0 to $N - 1$. This sum is easy to calculate if you remember the trigonometric formula for the cosine function of the double angle. It gives

$$2\cos^2(2\pi n/N) = 1 + \cos(4\pi n/N).$$

As the sum of the linear cosine terms is zero for $N \geq 3$ (prove this!), we find that the kinetic energy is

$$\frac{N}{2} \cdot \frac{1}{2} m\omega^2 A^2 = \frac{\lambda}{4} \rho_1 \omega^2 A^2,$$

where $\rho_1 = m/a$ is the linear mass density of the chain.

The result of the computation is simple and intuitively clear. The kinetic energy per unit length for the sinusoidal wave is proportional to the mass density of the chain and to the squared amplitude and frequency. Quite similarly, you can derive the potential energy of the springs.[78] The result is that the potential energy is equal to the kinetic energy, and thus the total energy per unit length (linear energy density) is simply

$$E = \rho_1 \omega^2 A^2 / 2.$$

In the continuum limit, a stronger statement is true: the kinetic energy is equal to the potential energy for any small part of the travelling sinusoidal wave at any moment.

Note that the normal modes (standing waves) behave like individual oscillators. In particular, the kinetic energy is not equal to the potential energy. Accordingly, the total energy per unit length of the standing wave is one-half of the energy of the travelling wave having the same amplitude and frequency (as an exercise, try to show this). To find the energy of an arbitrary wave, one has to know its Fourier expansion. The energy of the wave is the sum of the energies of its Fourier components.

Let us write the expression for the energy of any wave in the continuous limit, say for the elastic rod. The total energy of the wave stored in a small interval Δx is

$$\Delta E = \left(\frac{1}{2} \rho_1 \dot{y}^2 + \frac{1}{2} K y'^2 \right) \Delta x . \qquad (5.24)$$

I hope that the meaning of this formula is by now clear. The first term is kinetic energy and the second one is the potential energy. The total energy of a wave, a group of waves or of a soliton may be found by summing the energies of the small pieces given by (5.24). If the wave is in an external field (electric, gravitational, etc.) you add to ΔE the corresponding potential energy.

You may ask what the energy is of a travelling gravitational wave on the surface of water. For the rectilinear (or circular with a large enough radius) sinusoidal waves, the answer is very simple: the surface density of the total energy (the energy per unit area) is $(1/2)g\rho H^2$ where ρ is the volume density of water and H is the height of the wave. You may obtain this and the preceding formula by using dimensional analysis plus one additional assumption — independence of the energy density on the wavelength or its proportionality to the squared amplitude, whatever you prefer. As an exercise, try to find the energy density of the capillary and gravitation–capillary waves.

---

[78]First you have to calculate the sum of the terms $k(y_{n+1} - y_n)^2/2$ and then use the expression for the wave velocity $v = \lambda \nu$ in the chain, i.e., $v = a\sqrt{k/m}$.

Thus we see that the energy stored in any wave can be determined in a simple way, at least conceptually. Somewhat more difficult is the problem of energy transport. Probably the first solution for waves in elastic media was given by the Russian physicist N.A. Umov in 1874. However, his booklet *Equations of Motion for Energy in Solids* containing the notion of energy flow in waves and practical definitions of it was published in Odessa (in Russian) and was not known to the scientific community. An independent approach to energy transport by waves was later given in the paper of the English physicist and well-known expert in hydrodynamics, Osborne Reynolds (1842–1912): "On the Rate of Progression of Groups of Waves, and the rate at which Energy is Transmitted by Waves" (1877). Reynolds showed that the energy of the gravitational wave in water flows with the group velocity.[79] Even more importantly, he related energy transmission to the pressure of the wave. John Henri Pointing (1852–1914) applied these ideas to Maxwell's electromagnetic field (1884) and thus showed that waves of light also must produce a pressure, like a gas of particles. Thus the pressure of light on a mirror is equal to $p = u/3$, where u is the volume energy density of the reflected waves of light.

This result looked so bizarre to many physicists including the famous Lord Kelvin that they preferred to think that it demonstrated the failure of the Maxwell theory. The final resolution of all misunderstandings and even quarrels came only in 1901, with Lebedev's experiments that demonstrated the complete validity of Pointing's prediction. Today we know that light consists of photons and so existence of pressure is evident to us. However, in the wave picture of light this thing looks more difficult.

Mandelshtam discussed this difficulty in detail. The nature of the difficulty lies in that we are inclined to consider the *infinite* sinusoidal wave. According to Reynolds:

> This wave may be represented by a sequence of identical independent pendulums. The phases of oscillations of this pendulum may be so chosen that their oscillations will reproduce the shape of a progressive sinusoidal wave. However, there is no energy flow in this system ...

The difficulty disappears if we recall that the infinite wave is just a mathematical abstraction and that any physical wave is in fact a wave group. Any such group carries some energy through space, and the velocity of the energy transmission is the group velocity.[80] A dramatic consequence of this fact is that the energy and the phase of the wave may move in the opposite directions, if the group velocity happens to be negative. There is nothing paradoxical in such an effect. It only reminds us that phase velocity itself has little to do with energy flow.

---

[79] For deep water waves, the effect is especially dramatic as the group velocity is approximately half of the phase velocity.

[80] Hamilton called it the velocity with which Light overcomes Darkness.

There is no problem at all with soliton energy. The soliton behaves like a particle and its velocity and energy are well-defined. It also resembles wave groups but this resemblance is superficial. To demonstrate this once more, let us consider two colliding wave pulses on the D'Alembert string. At the moment $t = 0$, they are situated, say at the points $-x_0$ and x_0, and move with velocities v and $-v$, see Figure 5.11. At the moment $t = x_0/v$, they merge at the point O, and the shape of the resulting pulse is simply the algebraic sum of the two pulses. At the moment $t = 2x_0/v$, they have exchanged places and move in opposite directions.

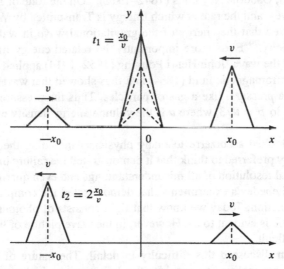

Figure 5.11 "Collision" of two wave pulses on the D'Alembert string–
the pulses are transparent to each other

This collision might seem similar at first to that of two solitons. However, the two pulses pass through each other without any interaction. Each pulse moves as if the other does not exist! In addition, the *shape of the pulses is completely arbitrary* while the velocity is fixed, it is equal to the wave velocity in the string. In contrast, the velocity of the soliton is arbitrary, but its shape is fixed (it may depend on the velocity but this is not important). Finally, the most important difference is that the pulse of a fixed shape exists only in an ideal string. Even tiny effects of dispersion and nonlinearity deform it and soon make it unrecognizable. On the contrary, Russell's solitons use nonlinearity to overcome the diffusing influence of the dispersion, and only friction gradually destroys them. But there also exist solitons which cannot be destroyed even by friction! The next chapter is devoted to these staunch solitons. I like them, very much!

Part III

The Present and Future
of the Soliton

What is now proved was once only imagined.

William Blake (1751–1827)

Chapter 6
Frenkel's Solitons

Let us return to more recent times and look at new concepts and accomplishments in soliton theory as introduced by theoretical physicists (theorists, for short).

What is Theoretical Physics?

> The important thing in science is not so much to obtain new facts
> as to discover new ways of thinking about them.
>
> William Lawrence Bragg (1890–1971)

Up to now we have been speaking about physicists or mathematicians. So who are these "theorists?" They are men/women engaged in theoretical physics. But then what is this theoretical physics? Ya. I. Frenkel once gave the following definition: "Physical theory is something like a suit sewed for Nature. Good theory is like a good suit.... Thus the theorist is like a tailor." On another occasion Frenkel formulated a more serious definition of this rare profession: "... a negative definition of the theorist refers to his inability to make physical experiments, while a positive one... implies his encyclopedic knowledge of physics combined with possessing enough mathematical armament. Depending on the ratio of these two components, the theorist may be nearer either to the experimentalist or to the mathematician. In the latter case, he is usually considered as a specialist in mathematical physics."

This division is not to be taken too literally. Nevertheless, we may say that the main task of the theorist is to interpret experimental facts by drawing "caricatures" and tailoring a "mathematical suit" for them. In other words, the theorist translates the vague, inarticulate language of Nature into the clear and precise language of mathematics. Once this goal is achieved, the theory is transferred to the department of mathematical physics for a final polishing and extraction of physical predictions and applications.

A no less important task of the theorist is to interrogate Nature, to ask new questions by suggesting new experiments. Clearly, you can ask a sensible question only if you have in mind some theory or, at least, a reasonable hypothesis.

> And Nature doesn't permit us to take
> the covers off Her beauty.
> By no machines you can force Her to tell
> the truth your mind hadn't guessed.
>
> Vladimir Soloviev (1853–1900)

So was summarized the relations between Nature and Man by a Russian philosopher and poet! By the way, poets are inclined to exaggerate the difficulties of human contacts with Nature: "indifferent nature" (Aleksandr Pushkin), "nature–sphinx" (Fyodor Tyutchev), and so on. Theorists are more optimistic: "Raffiniert ist der Herr Gott, aber boshaft ist er nicht," said one of the great theorists, Albert Einstein (most probably, he here identified God with Nature). Believing that Nature is understandable and, moreover, describable by laws of physics in mathematical language, the theorist first tries to explain new facts by using known concepts and theories. It happens, however, that this is impossible, and some unusual, "illogical" jump has to be made. In such a case, "encyclopedic knowledge" and "mathematical armament" may prove insufficient. The theorist has to use his imagination boldly to create new concepts having no analogy in the past. "Imagination is more important than knowledge" (A. Einstein).

> A friend of mine has defined the practical man as one who understands
> nothing of theory and the theorist as a dreamer who understands nothing at
> all.
>
> Ludwig Boltzmann

The separation of theoretical from experimental physics became a reality near the end of the 19th century, and was marked by publication of the first lecture courses in theoretical physics by L. Boltzmann, H.A. Lorentz, and M. Planck. (I know these beautiful lectures very well. They have been used in the 20th century until well into the 1930s.) An earlier theorist, J.C. Maxwell, didn't think of himself as a pure theorist. Probably, he wasn't. In his *Treatise on Electricity and Magnetism*, pure mathematical theory peacefully coexists with detailed discussions of observations, experiments, and with what we now call electrical engineering. Among a few of the "chosen" who succeeded in understanding Maxwell's theory and in explaining it to other physicists were Helmholtz and Boltzmann. Boltzmann was perhaps the first well-known pure theorist.[81]

Theoretical physics in Russia had a late start. In fact, it began to develop only in the 20th century. The main contributions to this development were made by several scientists born in the last decade of the 19th century and by some younger bright theorists who started their careers soon after the Revolution. The most important figure in this process was Paul Ehrenfest (1880–1933), a former student of Boltzmann's. He lived in St. Petersburg from 1905 to 1912, where he organized the first seminar on theoretical physics to be held in Russia. In 1912, H.A. Lorentz invited Ehrenfest to become his successor to the Chair of Theoretical Physics in Leiden. Ehrenfest, being at the center of the main events in 20th century physics, expended a lot of effort to help young Russian theoretical physicists to rise to world level in a very unfavorable post-revolutionary environment. In St. Petersburg, he supported young leaders of the theorists, Yu.A. Krutkov and Ya.I. Frenkel; in Moscow he especially supported I.E. Tamm,[82] the leader of theorists grouped around his teacher L.I. Mandelshtam. Ehrenfest helped young Russian theorists to travel to Western centers of science and to get financial support for their research from Western foundations. With his help, Krutkov and Frenkel obtained stipends from the Rockefeller Foundation. Later, similar grants were given to V.A. Fock and L.D. Landau.

These young bright, talented theorists laid the foundation of theoretical physics in Russia. Their personalities, views and interests were all different, but they all were, in a good sense, obsessed with science. Frenkel published the first complete course on theoretical physics to appear in Russia, and these articles were internationally recognized. Frenkel and Fock were the first teachers of quantum mechanics for Russian physicists (I have in my possession *Wave Mechanics* of Frenkel (1933) and *Basic Quantum Mechanics* of Fock (1932), very good textbooks, although printed on poor paper). Later L.D. Landau and E.M. Lifschitz started to publish the now famous Landau Course on Theoretical Physics.

Efforts of an older generation of physicists, in particular, A.F. Ioffe, P.P. Lazarev and L.I. Mandelshtam, combined with the very active work of young theorists, allowed Russian theoretical physics to survive and progress in terrible times after the Revolution. Russian theorists started their work too late to participate in establishing the foundations of the new physics, quantum mechanics. This had been done by N. Bohr, W. Heisenberg, E. Schrödinger, P. Dirac, M. Born, W. Pauli, and others. However, young Russian theorists were very successful in extending the applications of quantum mechanics to new fields, especially to condensed

---

[81] He was not quite alien to practice. For instance, it is known that he had designed and built an electric sewing machine for his wife. On the other hand, a typical 20th century theorist, Ya.I. Frenkel, was at a loss as to how to replace a burned out electric safety plug in his apartment.

[82] Tamm (1898–1971) was a student for one year at Edinburgh University. He had to return to Moscow at the beginning of the First World War.

matter physics, and later to statistical physics and quantum electrodynamics.[83] Another characteristic feature of the Russian theorists of that time was that they usually worked in good contact with experimenters and engineers. For some reasons, which I do not quite understand, contact with mathematicians was not so good. There were brilliant pure and applied mathematicians in Moscow and Petersburg, but it seems that most of them disliked modern theoretical physics and quantum mechanics in particular.

Ya.I. Frenkel's Ideas

As a musician, by first tacts, recognizes Mozart, Beethoven or
Schubert, a mathematician, quite similarly, by first pages can,
recognize his Cauchy, Gauss, Jacobi or Helmholtz.

Ludwig Boltzmann

Yakov Il'ich Frenkel (1894–1952) occupied his own special place in the constellation of theoretical stars of his time. The most typical feature of Frenkel's style was introducing very simple and original physical models combined with the use of minimal mathematical tools. I have heard that L. Landau once said: "Fock reduces every problem to partial differential equations, I — to ordinary ones, but Frenkel reduces everything just to algebraic equations." Of course, this is a joke but it certainly contains some truth.

Every true theorist must be, in a sense, omniscient, equipped to solve any physics problem.[84] Frenkel had this quality in the extreme and he was really interested in everything! His apparent peculiarities — an affection towards models and analogies, very sharp observations, his inclination to explain not only laboratory experiments but also physical phenomena, which we meet in everyday life — are akin to typical features of natural philosophers of the 19th century. Like them, with the same childish curiosity, he always watched and tried to explain things of the surrounding world which we today usually pass by without even noticing.

His son, V.Ya. Frenkel recalls that the ideas of some of Frenkel's physical investigations arose from such seemingly trivial observations. Sometimes it happened at home, as he watched bubbles appearing and growing in boiling water or looked at strange evolutions of the surface of the rotating tea in his teacup. He also liked to watch waves and wavy structures of sand on sandbanks. Maybe, while looking

---

[83] Their contribution is reflected in terminology. To give some examples of their achievements at that time, we say "Landau's diamagnetism," "the Fock space," "Tamm's surface energy levels," "Frenkel's excitons," etc.

[84] Lev Landau always stressed this and required from his students a broad knowledge of physics. Accordingly, at the end of the 1930s he conceived of writing a complete course in modern theoretical physics. This grand project was completed only after his death by his collaborators. The last and tenth volume of the Landau course of theoretical physics appeared in 1978.

at defects, discontinuities in these structures, he started to think about *dislocations*. Many of his works originated also from observations of clouds, lightning, vortices and whirlpools. Some of his colleagues found his affection for "bubbles and drops" outdated and even ridiculous, especially in the era of quantum physics. However, we now know that many of his ideas proved to be enduring and fruitful. He had a gift of vision for seeing unexpected new things in seemingly familiar, well-known phenomena. He was well aware of his gift and consciously cultivated it: "In our search of laws of nature, we must learn to see not so much Old in New as New in Old."

Although the range of his scientific interests was unbelievably broad — from lightning to atomic nuclei — his most celebrated achievements were in the theory of solids and liquids. In these fields, he formulated several fundamental new ideas. According to Nevill Mott:[85] "Ya. Frenkel was one of the founders of the theory of solids.... Every student in England knows "Frenkel's defect".... This is one of the fundamental ideas of physics." This defect is an "empty place," a "hole" in the crystal lattice, which can move through it like a negatively charged particle (it is supposed that the ion of the lattice has somehow left its place). Frenkel first formulated this idea in his paper published in 1926: "Taking into account the mobility of the holes, we may regard them as peculiar 'negative atoms'."

This idea is really fundamental and fruitful. Two years later, Paul Dirac conceived of holes in a "sea" of filled electron states and thus introduced "antielectrons," later named positrons ("positive electrons"). The idea of "holes" plays a major role in the theory of semiconductors. But all this was difficult to foresee in 1926. The real significance and depth of this simple idea became clear only much later. No easier was the fate of the idea of dislocations, first introduced by Frenkel and Kontorova in 1938.

Atomic Model of the Moving Dislocation after Frenkel and Kontorova

> ... what is theory? An uninitiated sees just a mass of incomprehensible formulas... However, they are not its essence.
>
> Ludwig Boltzmann

The *Frenkel–Kontorova soliton (dislocation)* is a special sort of a defect in the crystalline structure of solids. To be more precise, we will study a model of such a defect in the simplest possible model of a crystal. Although this model is sketchy, it allows us to qualitatively understand many relevant properties of real solids. In this sense, it is a realistic sketch.

---

[85]Nobel Prize winner, Sir N. Mott, was for many years the director of the Cavendish Laboratory (Cambridge), after J. Maxwell, W. Rayleigh, J.J. Thomson, E. Rutherford and L. Bragg.

A limiting case of the *dislocation* is the Frenkel "hole" mentioned above. Although this local hole can move through the crystal, the energy cost of moving may be rather high, because the atoms (ions) are strongly attached to their places. Much easier is to set in motion a defect in which the defect is, so to speak, distributed over many interatomic distances. Dislocations are such "distributed," non local defects.

Let us construct a very simple model of the dislocation. Imagine a periodic sequence of hills and hollows, Figure 6.1a. At the bottom of each hollow rests a small point mass ball bound by an elastic spring. The balls impersonate atoms, the springs — forces acting between atoms in one layer. The atoms of a neighboring layer are shown in Figure 6.1a by small crosses; in our model, their positions are fixed. Their influence on the first layer is represented by constraining the movements of the atoms of the first layer to the periodic curve (hills and hollows) which is placed in a vertical gravitational field. Thus, the periodic curve visually represents the gravitational potential energy of a single ball (in some scale).

It is obvious that the springs do not allow the simple Frenkel defect to exist (when one of the hollows is empty, while all the balls are resting at the bottoms). A possible equilibrium state is depicted in Figure 6.1b, which gives an idea of the Frenkel–Kontorova dislocation. In Figure 6.1c, we have drawn a diagram of the shifts of the balls from their equilibrium positions (compare with Figure 5.1). The resulting curve resembles the graph for the asymptotic movement of the pendulum shown in Figure 4.10. We shall soon show that the word "resembles" may be replaced by "coincides," if the size of the dislocation is much larger than the interatomic distance a. In the limiting case of very soft springs, the dislocation reduces to the Frenkel hole.

In the dislocation presented in Figure 6.1b,c, the concentration of atoms near the dislocation center is less than in the undeformed state (so we have a "rarefaction"). In the dislocation of Figure 6.1d,e, the atoms are concentrated near the center ("condensation"). In this last case, the graph of the shifts corresponds to asymptotic movement in the opposite direction. With very soft springs, we would get a state in which one of the hollows contains one extra atom.

Frenkel's defects arise in pairs – the atom leaving a hollow jumps to a neighboring hollow. Thus, at the moment of the creation of the defect we have, so to speak, a "particle" (hole) and an "antiparticle" (extra atom). As they can move, they may separate from each other and live independently. If they meet, they may "annihilate." The same is true for dislocations. Let us call the *dislocation of rarefaction* "positive" or simply *dislocation* while the *dislocation of condensation* will be called "negative" or *antidislocation*.

A long dislocation (much longer than a) is mobile because small shifts of each atom do not require a noticeable energy supply. So, the dislocation in an ideal crystal freely moves without changing its shape. The term "ideal" here means that all balls and springs are identical and the structure of the hollows and hills is ideally periodic. If there is one hill higher than all others, the dislocations will be "repulsed" by it. A smaller hill will act on dislocations like an attractor. Similar effects may be caused by inhomogeneities in the chain of the balls (different masses or different springs).

It is not very difficult to understand that two dislocations repel each other, while dislocations are attracted by antidislocations. The reason for the attraction is evident: in the dislocation the springs are stretched, while in the antidislocation they are contracted. From this point of view, it must be quite evident why two antidislocations repel. It may be slightly more difficult to understand repelling of two dislocations but, in fact, there is a complete symmetry between the positive and negative dislocations.

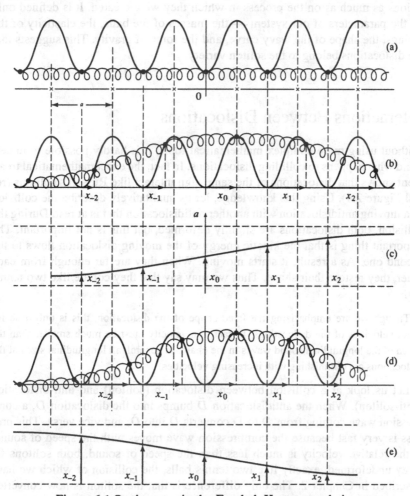

Figure 6.1 Static states in the Frenkel–Kontorova chain:
a) ground state– atoms are not displaced, b) dislocation, c) graph of displacements of atoms in the dislocation, d) and e) the same for antidislocation

What is not obvious and requires a mathematical analysis of the model is why the configurations of the atoms, which we call dislocations, may stay in equilibrium. It might happen that there is no such stable configuration. In the standard linear chain of atoms bound by springs, which we discussed in the preceding chapter, attempts to shift any of the atoms will invariably result in waves, and no stable configurations besides the trivial one will emerge. The same will be also true for the nonlinear chain, if the shifts of the atoms are small. However, for large enough shifts, it is possible to create the dislocation. Waves will also emerge but, when they run away (or die down due to friction), a stable dislocation may appear. Thus, any initial excitation gives rise to travelling waves, dislocation and antidislocation. The shape of the dislocations does not depend on the initial conditions as much as on the process in which they were created. It is defined only by the parameters of our system — the masses of the balls, the elasticity of the springs, the shape of the wavy curve, and the force of gravity. This suggests that the dislocations belong to the soliton variety.

Interactions Between Dislocations

Without using more complex mathematical tools, it is hardly possible to understand what happens to colliding dislocations. In fact, the exact mathematical treatment shows that dislocations of the same sign interact like Russell's solitons (recall Figure 2.4). Using this knowledge, let us qualitatively describe the collision of a moving antidislocation with another antidislocation that is at rest. During the collision, both dislocations are slightly deformed, but this is not important. The important thing is that the kinetic energy of the moving dislocation flows to the second one. As a result, it starts moving. When they are far enough from each other, they restore their shape. Thus we may say that they collide like two tennis balls.

Though we are emphasizing the fixed shape of the dislocation, this is only true for small velocities of the dislocations. In fact, the velocity must be much smaller than the speed of the longitudinal sound waves in the chain. In general, all longitudinal sizes of the dislocations are diminishing with increasing velocities.

Let us look at a collision between dislocation (soliton) and anti-dislocation (anti-soliton). When the antidislocation $\bar{D}$ bumps into the dislocation D, a compression wave coming from $\bar{D}$ to D converts D into $\bar{D}$, and vice versa. This process is very fast because the compression wave moves with the speed of sound. If the relative velocity is much less than the speed of sound, both solitons fly away undeformed, exactly like two tennis balls, the collision of which we have described in Chapter 2. The only difference is that the soliton is now converted into antisoliton and vice-versa. In this particular collision, it is quite obvious that the solitons do not penetrate through each other but rather collide like particles. The same is true but possibly not so evident for the D-D collision.

An interesting new feature of these solitons is that there exist two different species of them. They differ in their internal structure and may be labelled by a *charge*, which may assume two values: $+1$ for the dislocation D and -1 for the anti-dislocation $\bar{D}$. In the positive dislocation D, the overall effect of the displacement of the atoms is equivalent to that of Frenkel's hole, although it is spread over many interatomic distances. Accordingly, we ascribe to D the "charge" $+1$, while the charge of $\bar{D}$ must correspondingly be -1. These charges should be algebraically added, when we have several dislocations. For example, the total displacement of the atoms in the system of D and $\bar{D}$ is 0; it follows that its charge is 0. The state of N solitons has the charge N, and the system of N solitons plus M antisolitons has the total charge $N - M$. The conservation of this charge "explains" the stability of the states having a nonzero charge. This "explanation" simply gives a short formal expression of the topological nature of these solitons.

Let us recall that the above definition of charge is a matter of convention and convenience. This charge has nothing to do with the electric charges of atoms or electrons. Solitons are very similar to particles, but these particles exist only in a given lattice of atoms (or mass points with springs). Accordingly, their charges may be defined and are conserved only for solitons in the same lattice. In this sense, the lattice is a "universe" for these solitonic particles.

Nevertheless, this small one-dimensional "universe" is a very interesting laboratory for a particle physicist. We have particles and antiparticles, which are created and annihilated in pairs, like electrons and positrons in the real world. We have conserved charges resembling electric charges. We have also something like electromagnetic waves. It is not difficult to understand that these are small-amplitude travelling waves, in which the atoms oscillate near their equilibrium positions. In other words, they are just waves of sound in our lattice.[86] I like this small universe! In fact, it was this beautiful theory that attracted me to the theory of solitons.

Of course, this model of the world is too simple to be realistic. It is difficult to imagine that really complex systems can be created from these particles. The fact that this world has only one spatial dimension severely restricts possibilities for creating really complex structures (the fact that we, self-conscious systems, live in three dimensions is not accidental!). Nevertheless, this model gave rise to several interesting ideas in particle physics. At the end of this book, I'll try to describe several modern concepts of the particle world influenced by these simple solitons. The next section is devoted to a fairly complex and peculiar "atom" made of the soliton and antisoliton.

---

[86]Unlike the usual waves of sound, they have a strong dispersion for large values of the wavelength λ. Similar to deep water waves, their phase velocity v increases with λ. Later we will find the dispersion formula.

"Live" Solitonic Atom

It is not difficult to imagine that a dislocation and an antidislocation may, under certain conditions, form a bound pair resembling the hydrogen atom. To be precise, it is more similar to the so-called "positronium" atom, made of the electron and antielectron (positron). Unlike the hydrogen atom, positronium does not exist in nature, because its lifetime is very short. The electron and positron in positronium annihilate and convert into a pair of photons (the quanta of the electromagnetic waves), and so this atom lives only about $\sim 10^{-10}$sec. It might happen that the soliton and antisoliton also annihilate giving rise to waves of sound (the quanta of the waves of sound are called phonons but quantum effects are usually negligible in soliton physics). However, in the *continuous* Frenkel–Kontorova model there exist soliton-antisoliton atoms that might live indefinitely.[87] These atoms have a rather complex structure, but their main properties are easy to explain without much mathematics.

Let us imagine that a dislocation D and an antidislocation $\bar{D}$ are very close to each other, as shown in Figure 6.2. On the vertical axis, we have plotted the displacement of each atom divided by the interatomic distance a. On the horizontal axis, we also show the real position of the displaced atoms. The upper part of the figure shows the displacements when D is to the left of $\bar{D}$. Accordingly, in the left-hand side we have "rarefied" atoms, while in the right-hand side they are "condensed." To understand the processes that occur in this system, recall our description of the collision between D and $\bar{D}$, but now without kinetic energy, and without the whole system moving along the x-axis. Instead, the wave of condensation moves to the left and, at some moment, presented in the lower part of Figure 6.2, D turns into $\bar{D}$. Then the condensation wave starts moving to the right and the system returns to the initial state.

Thus we have a periodic process similar to a standing wave. However, it is a very strange standing wave. It is localized in space, but nothing keeps it on its boundaries! No linear equation can describe this thing. It is called a *breather* or, sometimes, a *bion* (meaning a "living particle"). Such strange creatures do not exist on the water surface and they are not easy to observe.

Above, we have described the breather at rest but it may move with a constant velocity. In collisions with other breathers or with solitons, it preserves its identity, and these collisions may be described as those of particles. To summarize, the breather is a true soliton that has properties both of particles and of waves. You cannot describe it as D and $\bar{D}$ tied together by a sort of a spring, because a standing wave is permanently "breathing" "inside" it. But neither can you say that this is just a peculiar standing wave because, as a whole, it behaves like an extended particle. You may say that it is both particle and wave. You may say that

---

[87]This statement is only applicable to idealized systems. Any physical system is not ideal (and not really continuous!), and the energy of the oscillations inside the solitonic atom will eventually dissipate.

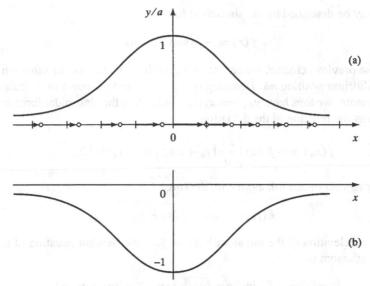

Figure 6.2 Configuration of displacements in the breather

it is neither particle nor wave. Both statements are true! If I remember correctly, Niels Bohr once (at least!) said that there are two sorts of truths. Normally, the opposite to a true statement is wrong. If both your statement and the opposite to it are true, you are saying a "deep truth"! In our case, the deep truth is that the solitons are really new objects of Nature.

At this point, I have to confess that, up to now, I was slightly cheating you. To make the description of dislocations more intuitive, I was using, instead of the Frenkel–Kontorova model, a somewhat different and more complex system. Now I am going to present you with the real theory of Frenkel–Kontorova dislocations. If you wish to get a true understanding of the theory of solitons, you must try to follow the mathematical arguments given in the next two sections. There, I essentially follow the original paper of Frenkel and Kontorova using, from time to time, the formulas of previous chapters that describe motions of the pendulum and waves in the linear chain. If you are not yet ready to exercise your mathematical abilities, you may skip all the mathematics and just have a look at the figures and the dispersion equations (6.9) and (6.10).

Dislocations and the Pendulum

In the FK model, the atoms move along a straight line (x-axis) and all forces are directed along the x-axis. The coupling between neighboring atoms in one layer may be represented by the springs, while the influence of the atoms of another layer may be described by the sinusoidal force

$$f(x) = -f_0 \sin(2\pi x/a).$$

As in the previous chapter, we denote by $y_n(t)$ the deviation of the nth atom from its equilibrium position na. Denoting by $x_n(t)$ the time-dependent coordinate of the nth atom, we thus have $y_n(t) = x_n(t) - na$. Then the sinusoidal force may be written as the function of the deviation

$$f(x_n) = -f_0 \sin\left[\frac{2\pi}{a}(y_n + na)\right] = -f_0 \sin\left(2\pi \frac{y_n}{a}\right).$$

The springs act on the nth atom with the force

$$k(y_{n+1} - y_n) - k(y_n - y_{n-1}).$$

As the acceleration of the nth atom is $\ddot{x}_n \equiv \ddot{y}_n$, the Newton equation of motion for the nth atom is

$$m\ddot{y}_n = -f_0 \sin(2\pi y_n/a) + k(y_{n+1} - 2y_n + y_{n-1}). \qquad (6.1)$$

With $f_0 = 0$, we would return to the linear equation (5.8), which we studied earlier. When $f_0 \neq 0$, the right-hand side of the equation is nonlinear in the coordinates of the atoms.

Equation (6.1) is the main equation of the Frenkel–Kontorova model. We will now find the solution of this equation, which describes one moving dislocation. If you have a good understanding of the previous chapter, you will not be confused by the fact that this is not a single equation but rather an infinite chain of equations. We know that the moving dislocation is something like a periodic wave in a chain of pendulums. Then, it is not difficult to understand that each atom exactly repeats the motions of the neighboring atom with some delay in time, Δt. This delay depends on the velocity of the dislocation v: $\Delta t = a/v$. Recalling Figure 5.7, we may thus write

$$y_{n+1}(t) = y_n(t - \Delta t), \quad y_{n-1}(t) = y_n(t + \Delta t).$$

Now, for a small enough time delay, the acceleration of the atom is approximately constant and we may apply (5.12):

$$y_n(t \pm \Delta t) \approx y_n(t) \pm \dot{y}_n(t)\Delta t + \ddot{y}_n(t)\frac{(\Delta t)^2}{2}.$$

Using this formula to also express y_{n-1} and y_{n+1} in terms of $\dot{y}_n$ and $\ddot{y}_n$ and inserting the result into the right-hand side of (6.1), we find the equation for y_n

$$[m - k(\Delta t)^2]\ddot{y}_n(t) = -f_0 \sin(2\pi y_n/a). \qquad (6.2)$$

Let us have a good look at this equation. You may see that it essentially coincides with the pendulum equation! On the other hand, when $f_0 = 0$, we return to the sound waves in the linear chain. In that case, (6.2) tells us that $m = k(\Delta t)^2$, and this gives the expression for the velocity of sound in the chain, $v_0 = a/\Delta t_0 = a\sqrt{k/m}$. Using this fact, let us introduce the notation

$$[m - k(\Delta t)^2] \equiv m\left(1 - \frac{v_0^2}{v^2}\right) \equiv -m_e . \tag{6.3}$$

When $v < v_0$, the so-defined effective mass m_e is positive.

Returning to equation (6.2), we see that it is the Newton equation for a "particle" of mass m_e moving in the periodic field of force. It is equivalent to the pendulum equation (4.1), and so let us introduce the notation that will explicitly show this. Recalling that $\sin(\phi + \pi) \equiv -\sin\phi$, we first define the angle variable ϕ_n by $2\pi(y_n/a) \equiv \phi_n + \pi$. This means that we measure the deviation of the nth atom from its equilibrium position by the "angle" ϕ_n. When the atom moves from its equilibrium position, $y_n = 0$ ($x_n = na$), to the extreme right of its "cell," $y_n = a$ ($x_n = (n + 1)a$), the angle increases from $-\pi$ to $+\pi$. Thus, when the atom moves between the boundaries of its "cell," our effective pendulum accomplishes the "asymptotic" movement corresponding to its separatrix (recall (4.9) and Figure 4.10).

To use what we know about the pendulum, we now rewrite (6.2) as

$$\ddot{\phi}_n = -\omega^2 \sin\phi_n , \quad \omega^2 \equiv 2\pi f_0/m_e a. \tag{6.4}$$

The asymptotic motion is defined by Eq.(4.9). Let us rewrite it in the form

$$\phi_n(t) = \pi - 4\arctan e^{-\omega(t-t_n)}.$$

Here the parameters t_n describe the time delay of the oscillations of the pendulums. To find these t_n, we have to recall that the dislocation (i.e., the wave of the asymptotic displacements of the atoms) moves with velocity $v = a/\Delta t$. It must then be clear that, to describe the dislocation wave, we have to choose $t_n = n\Delta t$. It follows that $\phi_{n+1}(t) = \phi_n(t - \Delta t)$ and thus $\phi_n(t) = \phi_0(t - n\Delta t)$.

In (6.2)–(6.4), we used discrete notation but, in fact, this notation is also valid in the continuum limit, when a is small. Let us formally pass to the continuum limit. With this aim, we express t_n in terms of the velocity of the dislocation, $t_n = na(\Delta t/a) = na/v$, and denote na by x. Then, $\phi_n(t) \equiv \phi(t, na) = \phi(t, x)$ and the dislocation may be written as

$$\phi(t, x) = \pi - 4\arctan e^{-\omega(t-x/v)} .$$

This function defines the shape of the dislocation at any moment t:

$$y_n(t) = a/2 + (a/2\pi)\phi(t, na).$$

It is instructive to rewrite the exponent in the form $(x - vt)/l_v$ where $l_v \equiv v/\omega$. Recalling now the definitions of "frequency" ω and of effective mass m_e (see equations (6.3) and (6.4)), you may easily find that

$$l_v = l_0 \sqrt{1 - \frac{v^2}{v_0^2}}, \qquad l_0 \equiv a \sqrt{\frac{m v_0^2}{2\pi f_0 a}}.$$

In the expression for l_0, we have the square root of the ratio of two energies. Let us find the meaning of these energies. The first energy $m v_0^2$ is simply $k a^2$, which is proportional to the energy increment of the spring made longer by a. In the denominator, we have the product of the force f_0 by the distance a; this is the energy required for shifting the atom to the neighboring cell when $k = 0$. To move the atom to the next cell, we have to work against two forces: the resistance of the elastic spring and the periodic force from the atoms of the neighboring layer. The size of the dislocation l_0 depends on the ratio of these two energies, and the formula for l_0 is simply a mathematical expression of the intuitive arguments presented above (recall Figure 6.1). When both energies are equal, $l_0 = a$ and we have Frenkel's defect. To get an extended dislocation, for which $l_0 \gg a$, the elasticity of the springs, ka, must be large as compared to the periodic force f_0, i.e., $ka^2 \gg f_0 a$. In what follows we assume this condition fulfilled.

Now, the final expression describing the dislocation is

$$\phi(t, x) = \pi - 4 \arctan e^{(x - vt)/l_v}. \tag{6.5}$$

This simple function is drawn in Figure 6.3b. In Figure 6.3a, we separately presented the dependence of this function on x at the time $t = 0$; then the center of the dislocation is at $x = 0$. Recalling that the atom at the point x is near an equilibrium position if ϕ is close to $-\pi$ or to π (for $\phi = -\pi$, the atom occupies its own equilibrium position) we see that the deviation from the equilibrium position is appreciable only close to the center of the dislocation, i.e., for $-l_v \le x \le l_v$. Thus l_v defines the *size of the moving dislocation*. For the dislocation at rest, the size, $l_v = l_0$, depends only on the lattice parameters. As we mentioned above, the size of the moving dislocation also depends on its velocity. The mathematical expression of this dependence,

$$l_v = l_0 \sqrt{1 - v^2/v_0^2}, \tag{6.6}$$

resembles the Lorentz transformation law of relativity theory. However, the relativistic formula contains, instead of the velocity of sound v_0, the universal constant of Nature, the speed of light. In addition, the meaning of the Lorentz transformation is conceptually different, and so we will not go into a more detailed description of this surprising analogy.

For those familiar with relativity theory, it might be of interest to know that the total energy E and the momentum p of the dislocation also have the "relativistic" form. One can show that their dependence on the velocity of the dislocation is

$$E = \frac{m_0 v_0^2}{\sqrt{1 - v^2/v_0^2}}, \quad p = \frac{m_0 v}{\sqrt{1 - v^2/v_0^2}}, \quad m_0 \equiv \frac{2}{\pi^2} \frac{a}{l_0} m. \tag{6.7}$$

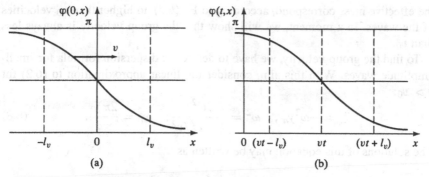

Figure 6.3 Moving Frenkel–Kontorova soliton

Thus, instead of Newton's mechanics, the dislocations obey the laws of relativistic mechanics. When the velocity of the dislocation is small enough, one may use standard Newton dynamics. Indeed, if $v^2/v_0^2 \ll 1$, the energy and the momentum of the dislocation are approximated by the well-known nonrelativistic formulas

$$E = m_0 v_0^2 + m_0 v^2/2 + \dots \quad p = m_0 v + \dots$$

I think that this analogy, as well as the model itself, would very much interest many scientists of the 19th century. For example, the English physicist Joseph Larmor (1857–1942) regarded electrons as a sort of dislocations in Ether. Technically, his theory was much more complicated than the FK model, but his basic concept of particles may be related to the soliton picture. Similar ideas attracted the attention of Henri Poincaré. In his very interesting report *New Mechanics* (1909) he said: "The inertia must be attributed to Ether rather than to Matter; only Ether resists motion, and thus we might say: there is no Matter, only the holes in Ether really exist." At the end of this book, the reader will find a sketch of modern ideas relating "elementary" particles to solitons of some nonlinear fields which, in a sense, play the role of the Ether.

The Fate of the Waves of Sound

Let us return to the FK model. We have examined dislocations that move like particles with velocities $v < v_0$. According to (6.3), this means that the effective mass m_e is positive. Now, instead of looking for the dislocation solution of this equation, let us try to find small-amplitude waves. For such waves, we may approximate the equation by the linear equation $\ddot{y}_n = \omega^2 y_n$. Here, ω^2 is positive for positive values of the effective mass. It must be clear that this equation has no oscillatory solutions. The reason is obvious: for positive ω^2, the force, $\omega^2 y_n$, drives the atom from its equilibrium position. On the contrary, when the effective mass is negative, we have the standard linear oscillator and thus there exists a harmonic wave. At first sight, this might look disturbing, because negative values of the effective mass correspond, according to Eq.(6.3), to higher than v_0 velocities of the waves. In a moment, we will show that the group velocity is always less than v_0.

To find the group velocity, we have to derive the dispersion formula for small-amplitude waves. With this aim, consider the linear approximation to (6.2) for $v > v_0$:

$$\ddot{y}_n = -\omega^2 y_n , \quad \omega^2 = \frac{\omega_0^2}{1 - v_0^2/v^2}, \quad \omega_0^2 = \frac{2\pi f_0}{ma}. \tag{6.8}$$

The solutions of this equation may be written as

$$y_n(t) = y_0 \sin\left[\omega(t - t_n)\right].$$

This expression describes a travelling wave if we choose, as above, $t_n = na/v$. Then, in the continuum limit, we have the standard sinusoidal wave

$$y(t, x) = y_0 \sin\left[\omega(t - x/v)\right].$$

The dependence of the circular frequency ω on the phase velocity v is given by (6.8). Recalling the standard relation $\lambda = v/\nu = 2\pi v/\omega$, we may find the dependence of the phase velocity on the wavelength λ:

$$v = v_0\sqrt{1 + \lambda^2/\lambda_0^2} , \quad \lambda_0 \equiv 2\pi v_0/\omega_0 = 2\pi l_0 . \tag{6.9}$$

Using the results of the previous chapter, you can find the group velocity u. The result is very simple $u = v_0^2/v$ (please try to derive it). You see that the group velocity for small amplitude waves is indeed smaller than the phase velocity and smaller than v_0.

It is instructive to write the dependence of circular frequency on the wavelength. This form of the dispersion law immediately follows from (6.9) if we replace v/v_0 by $\lambda\omega/\lambda_0\omega_0$:

$$\omega^2 = \omega_0^2(1 + \lambda_0^2/\lambda^2) . \tag{6.10}$$

The dependence of v/v_0 and ω/ω_0 on λ/λ_0 is schematically shown in Figure 6.4. You may easily deduce all properties of the dispersion of small amplitude waves from this picture.

The first remarkable property of the dispersion law in the FK chain is that the frequency of the travelling linear waves is always higher than ω_0, which is the frequency of the small amplitude oscillations of the individual atoms disconnected from the neighboring atoms. When the wavelength is very large, the distances between neighbors do not change and thus the chain oscillates undeformed. This means that the wave doesn't feel the presence of the springs; thus the frequency is defined solely by the atoms of the neighboring layer and must be close to ω_0.

The second important property of the dispersion law is easiest to see from the dependence of velocity on wavelength. The velocity $v(\lambda)$ is increasing with λ,

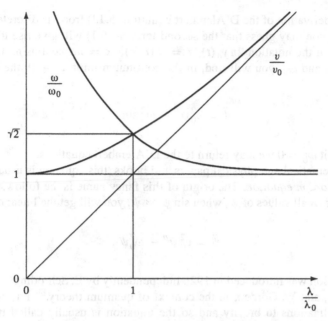

Figure 6.4 The dispersion law in the Frenkel–Kontorova chain
equations (6.9) and (6.10)

and the dependence is approximately linear for long waves (when $\lambda \gg \lambda_0$). This behavior is qualitatively similar to the dependence of the velocity of deep water waves on the wavelength (see (5.21)) and very different from the Cauchy dispersion law (see (5.17). Though for the waves on water $v(\lambda) \sim \sqrt{\lambda}$ while for the long FK waves $v(\lambda) \sim \lambda$, the nature of the dispersion is similar in both cases. Most important is that the dispersion law (6.9) or (6.10) has nothing to do with the discrete structure of the lattice and is defined by the action of the neighboring layer of the atoms. The role of this layer for waves in water is played by the gravity force.

By the way, in obtaining the dispersion law, we have neglected the discrete structure from the very beginning, supposing that $a \ll \lambda$ and $a \ll \lambda_0$. As $\lambda_0 = 2\pi l_0$, this also means that $a \ll \lambda_0$. Thus the dimension of the dislocation must be much larger than the interatomic distance a. It follows that Frenkel's defect, having dimension a, is beyond the scope of our theory. However, without insisting on a precise description, we may still regard the Frenkel defect as a dislocation, the dimension of which is comparable to the interatomic distance. A more correct description of the defect may be obtained from the original equation (6.1). If the springs are very soft, so that $ka \ll f_0$, there exists a stable state in which one atom is displaced close to the neighbor (say, $y_m = a - \epsilon_m$ for some m), while all others are close to their equilibrium positions (i.e., $y_n = \epsilon_n$, $n \neq m$).

I think that an attentive and thoughtful reader (having enough mathematical equipment) will find it not difficult to derive the continuum limit of (6.1). If you recall the derivation of the D'Alembert equation (5.13) from its discrete counterpart (5.8), you may guess that the second term in (6.1) will give rise to the term ka^2y''. With the notation $2\pi y_n(t)/a = \psi(t, x)$, $x = na$, and using the definitions of v_0 and ω_0, you will find, in the continuum limit $a \to 0$, the nonlinear equation

$$\ddot{\psi} = v_0^2 \psi'' - \omega_0^2 \sin \psi .\qquad (6.11)$$

Putting in it $\omega_0^2 = 0$ we may return to the D'Alembert equations.

In mathematical and physical papers and books, this equation is usually called the *sine-Gordon equation*. The origin of this fancy name is the following. If you write it for small values of ψ, when $\sin \psi \approx \psi$, you will get the linear equation

$$\ddot{\psi} = v_0^2 \psi'' - \omega_0^2 \psi .\qquad (6.12)$$

This equation was introduced in 1926 independently by E. Schrödinger, O. Klein, V.A. Fock, and W. Gordon, in the context of quantum theory.[88] Physicists have strong inclinations to brevity and so the equation is usually called the *Klein–Gordon equation*. A deformation of this name produced the strange hybrid "sine-Gordon."

At this point, an experienced reader would rightly guess that equations like these must have a record in the files of the history of science. Indeed, equation (6.12) was known in the 19th century as the equation describing motions of the string in an elastic medium (the second term in the right-hand side of the equation represents the local action of the medium on the string). The "sine-Gordon" equation was known to several mathematicians by the end of the 19th century (they couldn't imagine its later use in physics, of course). The appearance of this equation may even be traced to early investigations in Lobachevskii geometry,[89] but it was familiar to few geometers. A detailed study of the sine-Gordon equation was attempted as late as 1936. Then a German mathematician R. Steuerwald derived its solutions corresponding to one soliton, two solitons, and the breather (in modern language). These results were known to few experts in geometry and so Steuerwald's results had no influence on the development of soliton science. In physics, the sine-Gordon equation had been introduced by Frenkel and Kontorova quite independently of all this mathematical development. They had found its soliton solution, and so it is natural to call the soliton and the equation by their names. Most striking properties of this soliton were discovered much later by other scientists, but the first steps were taken by Frenkel and Kontorova.

---

[88] In fact, they considered a generalization of this equation to three space dimensions when ψ depends on t, x, y, z.

[89] Each solution of the equation defines a surface on which the axioms of Lobachevskii geometry are satisfied. These concrete realizations had played a major role in accepting Lobachevskii's and Bolyai's ideas by the mathematics community.

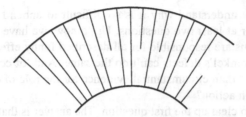

Figure 6.5 A schematic representation of dislocations in a bent rod

Let Us Have a Look at Dislocations

Everyone knows that pure metals are fairly soft (plastic). Of course, this is true for "ideal" metals, having a simple crystal structure not distorted by foreign impurities or internal defects. The crystal lattice of most common metals is constructed from equal "cubes." This cubic structure allows easy slips along certain planes (with a small applied force). For such deformations, you need not simultaneously shift all the atoms in the plane, because the applied shear force creates dislocations that may travel along the crystal without resistance. Thus, dislocations make pure metals with regular lattice very soft.

However, you might happen to observe that two seemingly identical pieces of wire made of the same metal (e.g., copper) may require different forces to bend them. The explanation of this strange behavior is that the internal structure of these pieces is different because they have different history. Most probably, those that are difficult to bend have been bent and unbent prior to your experiment. You possibly know that it is more difficult to unbend wire than to bend it. The explanation of this phenomenon is that, in the process of bending, a large number of dislocations is created inside the crystal lattice. The distances between the atoms in the crystal in this process are not changed, but some parts of the lattice are stretched, while other parts are contracted. Very schematically, the effect of bending is shown in Figure 6.5, where the cut-offs of the regular atomic chains correspond to dislocations.

When there are many dislocations, the crystal is so imperfect that the free motion of dislocations becomes difficult. In terms of our simplified model shown in Figure 6.1, this means that, instead of the periodic structure of bottoms and hollows, there emerges a very irregular sequence of hills and ravines that can catch the dislocations. Thus, to unbend the wire, you have to shift many atoms instead of "redislocating" small groups of them. By the way, forging metals also creates many interlocked dislocations, and this explains strengthening of metals too. On the contrary, in heated metal, dislocations disappear and the metal may return to the original ideal state. Thus, blacksmiths have used dislocations for hundreds of years in their art. You may puzzle your less educated friends by a very simple experiment. Take a soft thick copper wire, or make it soft by heating it in a flame (it remains soft after cooling). Then you may easily make a ring and suggest that your trustful spectator unbend it!

I hope you now understand why it is so difficult to unbend the wire, but we have yet to answer at least two questions. Up to now, we have simply supposed that the dislocations are responsible for all described funny effects. But you may rightly ask why Frenkel's defects can't do the same. And, of course, it would be better to have more than circumstantial evidence for the role of dislocations. Can we observe them in action?

Let us first try to clear up the first question. The answer is that creating a dislocation costs less energy than creating Frenkel's defect. Indeed, the energy of the defect at rest is approximately equal to ka^2, while the energy of the dislocation is

$$E_0 = m_0 v_0^2 = m v_0^2 \frac{2}{\pi^2} \frac{a}{l_0} = ka^2 \frac{2}{\pi^2} \frac{a}{l_0}.$$

It is given by (6.7) for $v = 0$. Now, it must be clear that the energy of the defect is much larger than the energy of the dislocation, because in real crystals the size of the dislocation, l_0, is much larger than the interatomic distance a.

The answer to the second question had been given by beautiful experiments done at the Cavendish laboratory about 40 years ago. Using the electron microscope, experimenters succeeded in seeing the real dislocations in crystals. Moreover, they had made a movie of dislocations that, using their description, "were fussing like mice."

You may see the dislocation pattern in a simple model of two-dimensional crystals invented by L. Bragg and G. Nye. These English scientists made a crystal-like structure from soap bubbles. Usually, any specimen of such a crystal contains several dislocations at rest, and they are easy to see. I will not go into a detailed description of the experiments with this bubble-lattice but rather recommend that you read their paper (see Appendix to Chapter 30 of the *Feynman Lectures in Physics*). The paper is easy and nice to read, and in Feynman's Lectures you may find other interesting remarks about dislocations.

Desktop Solitons

You can make a very simple device for observing basic properties of the FK solitons by once more using the chain of clips on a rubber band, mentioned in Chapter 5; see Figure 5.2. Just attach to the ends of the clips small pieces of plastic and the apparatus is ready! With a good (elastic) rubber band, this device is a chain of pendulums acting on each other through the rubber torsion. If you have carefully studied the models described above, you will not find it very difficult to write the equations for the pendulums and to predict their behavior.

To write the equations, it is advisable to idealize the device. Imagine that the clips are massless rods of length l, with the point masses m at their ends; then, the rubber band should be replaced by an ideally elastic string. This idealized model is schematically represented in Figure 6.6 (it should be remembered that the rods move in the planes perpendicular to the string and the gravity field acts on them in the vertical direction). When you

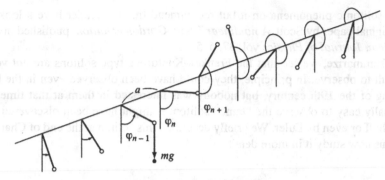

Figure 6.6 Desktop soliton

rotate the pendulums, the string is twisted and creates a torque, depending on the positions of the nearest neighbor pendulums. The string torque, acting on the nth pendulum, may be written as $K(\phi_{n+1} - \phi_n) - K(\phi_n - \phi_{n-1})$. The torque, created by the gravity force, is $-mgl \sin \phi_n$. Their algebraic sum must be equal to the product of the moment of inertia ml^2 by the angular acceleration $\ddot{\phi}_n$. Thus you will obtain an equation like (6.1). Repeating the above consideration you will find the soliton solution for this model. I leave all this as an exercise to the reader. I only mention that the "velocity of sound" in this device is $v_0 = a\sqrt{K/ml^2}$ and the size of the soliton at rest is $a\sqrt{K/mgl}$. You can measure these parameters and thus estimate K; other parameters are under your control.

Of course, this device is too primitive to study all the properties of the FK soliton. At most, you will succeed in seeing one soliton and observing the dispersion of the small amplitude waves. In Figure 6.6, we show the configuration of the pendulums corresponding to one soliton, when the angles change from 0 to 2π. It is easy to prepare the two-soliton state (the angles change by 4π), if your chain is long enough. However, in the rubber band model, it is practically impossible to observe the solitons in motion.

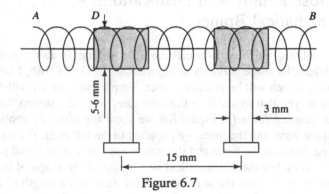

Figure 6.7

A much better device depicted in Figure 6.7, using the same idea, was designed by Alwyn C. Scott in 1969. I will not describe this device and the observation of many solitonic phenomena on it but recommend that the reader have a look at the original paper by Scott, *A nonlinear Klein–Gordon equation*, published in the *American Journal of Physics* vol. 37, p. 52.

To summarize, we see that the Frenkel–Kontorova type solitons are not very difficult to observe. In principle, they could have been observed even in the beginning of the 19th century, but nobody was interested in them at that time. It is equally easy to observe the "tame" soliton. It might have been discovered by Kirchhoff or even by Euler. We briefly described this soliton at the end of Chapter 4; let us now study it in more detail.

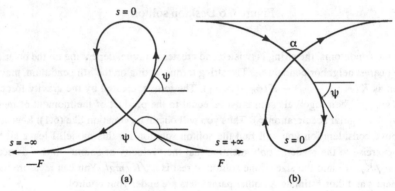

Figure 6.8 Euler's soliton (a) and antisoliton (b)

Other Close Relatives of Dislocations:
the Mathematical Branch

The "tame" soliton, which we will call the Euler soliton, is also closely related to the pendulum. Its shape is drawn in Figure 6.8a. In Figure 6.8b, I have drawn the antisoliton, which will be discussed later. Simple experiments with thin steel wire will show you that to stabilize this structure, you must stretch the wire by applying an external force (in Figure 6.8 we show the force F, applied to the right end of the wire, and the force $-F$, applied to the left end). The equilibrium condition requires that the force and the torque applied to each small part of the wire must be zero. For the mathematical description of the shape of the soliton, it is convenient to consider the angle ψ as a function of the length of the curve s. Then the condition of equilibrium for a small piece of wire of length Δs may be written as an equation for $\psi(s)$. Indeed, the momentum of the external force is approximately equal to $F \Delta s \sin \psi(s)$. It can only be compensated for by the momentum, generated by bending of the wire. This is more difficult to represent in mathematical terms.[90]

It is not difficult to guess that the torque must be proportional to the first or to the second derivative of $\psi(s)$. Of course, the simplest possibility is the first derivative $\psi'(s)$. However, you may easily see that this hypothesis doesn't work. Indeed, let us inspect Figure 6.9, in which the solid lines are graphs of $\psi(s)$, $\psi'(s)$, and $\psi''(s)$ (the dashed lines represent the antisoliton).

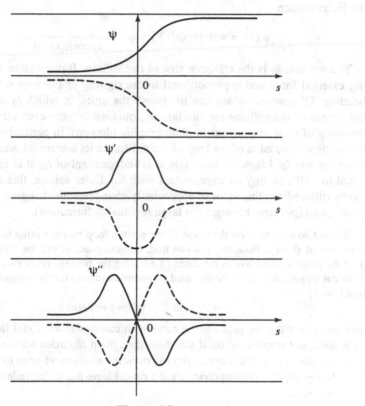

Figure 6.9

Now the maximum torque of the external force F corresponds to $\psi = \pi/2$, while the torque is zero for $\psi = \pi$. Moreover, the torque changes sign when we move from the negative values of s to the positive ones. Evidently, this is incompatible with the behavior of ψ', and qualitatively agrees with the dependence of the second derivative on s. Thus, the equation, describing the shape of the wire, may be

---

[90]To the meticulous reader, I recommend constructing a discrete model of the thin wire from small rigid rods tied into a chain by hinges with springs resistant to bending. Then you may show that, for the nth rod, the momentum of the force, generated by the spring hinges, is $K(\psi_{n+1} - \psi_n) - K(\psi_n - \psi_{n-1})$. In the continuum limit, this will give the second derivative in s.

written as

$$\psi'' = \frac{1}{l_0^2} \sin \psi(s),$$

where the parameter l_0 has the dimension of length. This equation practically coincides with the pendulum equation (4.1). In fact, the angle $\phi(s) \equiv \psi(s) - \pi$ satisfies the equation (4.1) with ω_0 replaced by l_0. Using the soliton solutions of the pendulum equation, we may immediately find the dependence of ψ on s for the Euler soliton

$$\psi(s) = 4 \arctan(e^{s/l_0}), \quad \psi' = \frac{2}{l_0 \cosh(s/l_0)}.$$

You see that l_0 is the effective size of the soliton. It diminishes with increasing external force and is proportional to the rigidity of the wire with respect to bending. Of course, we are free to choose the units, in which $l_0 = 1$, because the shapes of the solitons are similar, i.e., obtainable from each other by simple rescaling of their dimensions (a photographic blowup). In particular, the ratio of the vertical dimension of the loop of the soliton to its horizontal dimension is the same for any l_0. Likewise, the angle α is independent of l_0; it is approximately equal to $110°$. If, in your experiments with the Euler soliton, this angle significantly differs from this value or appreciably changes with changing F, your wire is not good (perhaps, having a too large residual deformation).

It is not so easy to draw the shape of the soliton loop by only using the obtained dependence of ψ on s. However, you can find, without much effort, the coordinates x and y of the point on the curve as functions of s. Using the obvious relations $y' = \sin \psi$ and $x' = \cos \psi$, you may derive the so-called parametric equation for the solitonic curve (recall that $l_0 = 1$)

$$y(s) = 2/\cosh s, \quad x(s) = s - 2 \tanh s .$$

You may check that the points of the curve with coordinates $x(s), y(s)$ lie on the circle of radius 2 and center at $(s, 0)$. If you draw the system of circles with centers at points $(na, 0)$, where a is a small number, you can draw the whole curve point by point, starting from the top, $(0, 2)$, and using dividers and a ruler. I leave this to the reader as an exercise; see Figure 6.10.

Up to now we only considered the Euler soliton at rest. Let us try to understand whether it can move. In experiments, it usually stays in the middle of the wire if the wire is good enough. This very fact shows that it can move; otherwise it would rest anywhere! To understand this simple effect, you have to recall that your wire has a finite length. Then the ends of the wire repel the soliton, and the equilibrium is obviously achieved in the midpoint. In a long wire, the soliton may stop at some other point because your wire is never ideal; it may have residual deformations obstructing its motion. If the wire were ideal and very long, in comparison to the soliton size l_0, you could see the motion of the soliton. Then you might also find that two solitons repel while soliton and antisoliton attract each other.

The important property of the Euler soliton is its "indestructibility." In an infinitely long wire, you cannot destroy or create one soliton. Of course, we tacitly

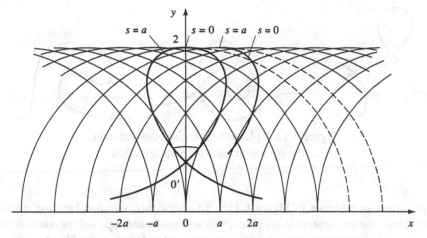

Figure 6.10 A graphical construction of the Euler soliton

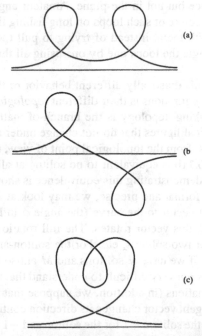

Figure 6.11 Two solitons (a); soliton-antisoliton pair (b);
a planar deformation of the two soliton configuration (c)

assume that the wire always remains in the plane; in three-dimensional space you
can make any loops of this sort (however, it is impossible to make some knots
even in three-dimensional space). Likewise, you cannot make two solitons (Figure 6.11a) if the wire remains in the plane. However, it is easy to create a soliton-

Figure 6.12 A planar topological deformation
of the soliton-antisoliton configuration

antisoliton pair shown in Figure 6.11b. To check this, you had best use a soft cord; by slow movements in the plane, you will easily untangle soliton-antisoliton loops. If you try to similarly disentangle two soliton loops, you will get the loop structure shown in Figure 6.11c. This double loop may be easily untangled in three-dimensional space but not in the plane. Amateur anglers often have problems with complex structure of such loops on long fishing lines. An expert angler knows how to deal with them: instead of trying to pull the string in rapidly, he will carefully disentangle the loops one by one using all three dimensions of our space.

A deep reason for this drastically different behavior of the soliton-soliton and soliton-antisoliton configurations is their different *topological* structure. I remind you that, loosely speaking, topology is the branch of mathematics studying the properties of geometrical figures that do not change under any continuous deformations of them. Thus, from the topological point of view, the soliton-antisoliton configuration Figure 6.11b is equivalent to no soliton at all. A sequence of continuous deformations demonstrating this equivalence is shown in Figure 6.12.

To make this more formal and precise, we may look at the angle between the x-axis and the tangent vector to the curve (the angle ϕ in Figure 6.8). When we move along the curve, this vector rotates. The full rotation of this vector is 2π for one soliton, 4π for two solitons, etc. For the soliton-antisoliton pair the full rotation is zero. Thus, if we have N solitons and M antisolitons, the full rotation is equal to $N - M$. It is not very difficult to understand that this number conserves for continuous deformations (in addition, we suppose that the curves are always smooth, that is the tangent vector changes its direction continuously). So, we may define the "charge" of the soliton: $+1$ for the soliton and -1 for the antisoliton; the charge of any collection of solitons and antisolitons is simply the algebraic sum of the individual charges. Due to its conservation under topological deformations, it is called the *topological charge*. Solitons with a topological charge are called *topological solitons*. The FK soliton is a typical topological soliton (it is clear that, in the infinite chain, the atoms at infinity must occupy their equilibrium positions; otherwise the energy of the configuration will be infinite). The topological nature of the soliton in the Scott model is even more obvious. There exist other topological solitons. Very interesting topological solitons exist in superconductors, to be discussed in the last chapter. In the next section, we will make acquaintance with topological solitons in magnetic phenomena.

Magnetic Solitons

Solitons similar to dislocations may exist in many other physical systems. Some of them will be briefly reviewed at the very end of this book. Here, I only present a sketch of magnetic solitons, which are interesting relatives of dislocations, and were first studied quite independently of them.

To describe magnetic solitons, let us concentrate on the most typical magnetics, like iron (*Ferrum*). All similar magnetic materials are generically called *ferromagnetics*. The main property of ferromagnetics is that they may be easily magnetized by applying an external magnetic field; then they produce their own magnetic field, i.e., become magnetic. To understand the role of magnetic solitons in the process of magnetization, consider a very simple model. Let us imagine that the pendulums in Figure 6.6 are replaced by small magnets (we may think of them as arrows directed from the South to the North pole of the magnets). We also replace the field of gravity by a magnetic field, which imitates the magnetic field existing in the crystalline lattice. This "crystalline field" makes the energy of each individual magnet minimal in its vertical position, i.e., when $\phi = 0$ or $\phi = \pm\pi$. It is not difficult to find that the simple dependence of the crystalline magnetic field on ϕ, $-\sin 2\phi$ satisfies this condition.

If there were no interaction between neighboring magnets of the chain, it would not have any of its own magnetization. On average, the directions of our small magnets will alternate, and thus they will not create any magnetic field around the chain. If we apply an external magnetic field, the chain will be magnetized, but after switching off the external field, this "induced" magnetization will also disappear. To keep the magnetization, additional forces between the neighboring magnets are needed (like those created by the rubber or spring torsion in the pendulum chain). These forces might ensure that the magnets hold parallel positions (say, the North pole is up) even in the absence of the external magnetic field. The nature of these forces in real ferromagnetics became clear only after the creation of Quantum Mechanics. The basic ideas on these quantum forces were introduced by Ya. I. Frenkel and Werner Heisenberg (1901–1976). Heisenberg's mathematical model describing the interaction of small molecular magnets in ferromagnetics is very profound and beautiful; it still reveals its secrets to new generations of theorists. For a qualitative understanding of magnetic solitons, we need not look into details of so-called "exchange" forces between molecular magnets. It is sufficient to know that they act like springs connecting the pendulums in the Scott model mentioned above.

Exchange forces make individual magnets rather collectivistic creatures. If, say, the left and right neighbors of a magnet are looking up, then it will also look up. Thus, molecular magnets in the ferromagnetic are strong conformists, they like to be ordered. In the language of physics, we say that to have the same directions of the magnets is energetically favorable (the energy of such a configuration is the lowest one).

We, however, forgot about normal magnetic interaction between molecular magnets, which act against the "conforming-type" ordering of exchange forces. As distinct from quantum exchange forces that really act among the nearest neighbors, these well-known classical forces have a long range. Accordingly, it becomes energetically favorable for the collection of the magnets to divide into many groups. In one half of these groups, the magnets look up, in the other half, they look down. These groups are called *domains*. In real crystals the division into domains may be more complex, but this is not essential for our crude picture.

The borders between domains are usually called *domain walls*. We also call them *magnetic solitons* because they are very similar to dislocations and mechanical solitons considered above. It must be clear that the magnets do not change direction suddenly and the thickness of the wall (the soliton dimension) depends on the parameters of the magnetic crystal. Like dislocations, the domain walls may move along the crystal if there is no obstruction from the crystal imperfections or from other domain walls.

If a ferromagnetic specimen is not magnetized (say, after demagnetizing it by annealing), it consists of many domains having opposite directions of their magnetic "collectives." In an external magnetic field, the domain walls start moving. When they stop, the dimensions of the domains, in which the molecular magnets are parallel to the magnetic field, become larger than those of the domains with oppositely directed magnets. When we switch off the external field, the walls start returning to their original positions. However, in any real, imperfect crystal, there is no reversibility. The domain walls will be stopped by imperfections, and oppositely-oriented domains will survive. Thus, our specimen becomes magnetized, and from now on, it will behave as a magnet. This mechanism of magnetization is so similar to that of the plastic deformations that the terms (magnetically) "soft" and "rigid" iron need no explanation.

The domain walls are somewhat easier to observe than the dislocations. A typical distance between two domain walls is about 10^{-3}cm, and their thickness is typically 10^{-5}cm or so.[91] Thus, it must be not too difficult to make the domains and domain walls visible. Let us prepare a small ferromagnetic brick, with well-polished faces, and suppose that the molecular magnets inside the domains are parallel to the upper face, as shown in Figure 6.13. Then the magnetic field of the internal magnets will not extend out of the upper face except at the position of the domain walls. If you now cover this face with a thin powder of a magnetic material, its particles will gather near the domain walls. Experiments of this sort were performed in 1931 by the American physicist Francis Bittner (1902–1967).

---

[91] As these dimensions are much larger than the interatomic distances, which are 10^{-8}cm, one may neglect the atomic structure and use a continuous approximation. It also follows that very small particles of a ferromagnetic material, of the size 10^{-4}cm or less, cannot have domain walls inside.

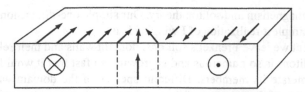

Figure 6.13 A schematic picture of the domain wall in ferromagnetics

At that time, domain walls had a record of theoretical investigation. The idea of domains in ferromagnetics was put forward in 1907 by the French physicist Pierre Weiss (1865–1940). In 1930, Ya. I. Frenkel and Ya. G. Dorfman explained why the domain structure of ferromagnetics is energetically favorable. They also estimated the domain size. The American physicist Felix Bloch, among other achievements, estimated the thickness of the domain walls. In 1935, L.D. Landau and E.M. Lifschitz developed a general and precise theory of domains and domain walls. In addition to a precise mathematical description of the internal structure of the domain wall, they found the equation, describing its motion under influence of the external magnetic field. Their equations are still used in soliton theory.

In spite of all these successes, the solitonic nature of the domain walls became clear much later. The Landau–Lifschitz equations are even more complex than those of Frenkel–Kontorova, so no wonder that for many years it did not occur to anyone that the domain walls are particle-like objects similar to dislocations. Of course, before creating the general theory of solitons, it was quite impossible to see any relation between domain walls and KdV solitons.

Recalling stories of other solitons, you might guess (correctly) that domain walls should have been observed in the 19th century. Indeed, about 100 years ago the Scottish physicist James Ewing (1855–1935) made a model in which he had seen magnetic domains and domain walls. The idea of his model goes back to the beginning of the 19th century. It was Fresnel who first proposed the idea of molecular magnets, in his letter to Ampère. Ampère developed this idea and mathematically studied the model of a gas with magnetic molecules. For this reason, molecular magnets were often called Ampère's molecular magnets. Later, W. Weber suggested making these molecular magnets pendulums. Thus emerged the lattice of magnetic pendulums.

Starting from this, Ewing introduced the idea of interaction between neighboring magnets and constructed the first consistent theory of ferromagnetism. For us, it is most interesting that he experimented with the lattice of small compass needles coupled to nearest neighbors. In this model, very similar to the much later model of Bragg and Nay, he tried to check his theory. Among other things, he observed the domains and even movements of their boundaries.

Ewing's theory and experiments (not only with the model but also with real magnetics) strongly stimulated P. Weiss but later were forgotten. In our quantum times, his theory is surely obsolete. However, his lattice model is still interesting for teaching magnetism and soliton theory. Our simple one-dimensional magnetic chain is the simplest realization of Ewing's ideas.

At this point we leave Frenkel's solitons, domain walls and their relatives. This family of solitons is so numerous and is growing so fast that it would be difficult even to enumerate its members. Different species of the domain walls may be met on all levels of the structure of the universe, from extremely small "walls" in elementary particle theory to extremely large ones in the theory of the creation of the universe. Possibly, you will meet this fruitful concept also in other branches of modern science.

Chapter 7
Rebirth of the Soliton

Someone will find footsteps of you
And take your course, span by span...

Boris Pasternak (1890–1960)

If our descendants try to comprehend what was most important in science and technology in our times, they probably will put computers in first place. Now the wildest dreams of Babbage are reality, and Maxwell or Kelvin would not have to spend time and efforts on dull calculations that could be executed by computers. Moreover, modern computers can do even some creative work, of which these great men could not even dream.

Babbage, Kelvin and Maxwell thought mainly of calculations with computers: numerical integration, solving differential equations, calculating complicated functions, and all that. Modern computers easily cope with all these tasks, and with speed unimaginable in the last century, but these are among the simplest applications of computers. More important today is using them for *symbolic calculations*.

In fact, we can teach computers algebra, analysis, differential equations, and so on. Naturally, all mathematical problems mentioned in the preceding chapters are easy exercises for them. They will solve our equations, giving us analytical expressions that describe the shape of the soliton, draw pictures and even make movies of motions of several solitons. Of course, all of this is possible for simple models only. In more complicated situations, we still use computers in the old-fashioned manner for numerical calculations; but the capabilities of modern computers are such that this work is completed quickly and the results are presented in a very comprehensible form. Instead of huge tables of numbers, the

A.T. Filippov, *The Versatile Soliton*, Modern Birkhäuser Classics,
DOI 10.1007/978-0-8176-4974-6_7, © Springer Science+Business Media, LLC 2010

computer provides us with all sorts of visual information and, what is most important, one can work in the *interactive mode*, i.e., changing strategy with our growing understanding of the problem.

These new possibilities are fundamentally changing the style of work in theoretical physics and applied mathematics. In the first place, the very notion of what it means to "solve" a problem is changing. Say we wish to study motions of two small massive bodies bound by a spring. To do this, we may simply write equations (5.1) and leave the rest to the computer. For the computer, it is not difficult to solve similar problems for five, ten, or one hundred bodies, and it will do this much faster than we can.

Then why have we spent so much time studying those tiresome derivations in Chapter 5? Was it for nothing? Of course, not! For us, the "body-spring" model was not the end in itself. Our aim was rather to gain an understanding of much more complex systems. In addition, our interest was focused not on particular movements of particular bodies but on a qualitative behavior of the system as a whole. We tried to find the regularities in these movements that would give a clear picture of all physical phenomena in this and other similar systems. By studying our simple model, we have prepared ourselves to consider more complex things. In this way, our imagination, intuition, and even emotions have become alert. Unfortunately, computers are devoid of these human qualities. So, there must be a sort of division of labor between men and computers. Man is the boss, while his computer is a servant, a very effective one but still lacking most important human qualities.

In solving scientific problems and, especially, in raising new problems and creating new concepts, the personality of the scientist plays a vital role. At first glance, this statement might look somewhat strange, as science is created by the collective efforts of many scientists and so must be impersonal. Then, why are human qualities of scientists of any importance in their work? Well, the final results of scientific research, checked in experiments and embodied in technical devices, are completely objective and impersonal. However, before becoming accessible to other scientists and to society, any scientific idea must be born in an individual's mind, and so creative work is a very intimate thing. For this reason, the computer is not likely to become an independent creative thinker, and yet, a cooperative work of creative scientists with computers may fundamentally change all aspects of scientific work. In this, the computer may become not only an obedient servant but also a powerful friend.

Can Men be on Friendly Terms with the Computer?

We are in the midst of a computational revolution that will change science and society as dramatically as did the agricultural and industrial revolution.

Norman J. Zabusky

It is too late to ask this question as the answer is too obvious. This friendship is a fact of our everyday life. Today, many physicists, especially the young, cannot even imagine their lives without computers. I am not speaking of everyday routine work, like preparing a manuscript on a PC, or carrying out tedious computations, and so on. This friendship is much deeper, often so deep that some physicists are even depending on the computer to help achieve a better understanding their favorite ideas. So, it is more than a friendship, maybe a "romance". Is there any reason for this, if the computer is simply an honest servant making complicated calculations for his boss? What is so attractive about the computer?

Well, doing calculations that are otherwise unfeasible is a very nice feature of the computer but not a reason for passion. More important is that the computer can suggest and help us to realize new approaches to our mathematical models of the world. It really can help us to discover new facts and invent new ideas. The computer doesn't know amazement or admiration, but it can astound or ravish us! This idea was alien even to the shrewdest thinkers of the 19th century. In her description of the analytical machine suggested by Babbage, Lady Lovelace stressed that the machine can't pretend to invent anything really new. Its destiny is just to follow our prescriptions. Babbage certainly held the same opinion, as did Maxwell and Kelvin.

John von Neumann (1903–1957) and Alan Turing (1912–1954), [92] founders of the modern theory of computers, had a radically different opinion of this matter. In his unpublished notes "Intelligent Machinery", written in 1947 and later expanded to the widely known paper "Computer Machinery and Intelligence" (1950) (also known as "Can a Machine Think?", Turing wrote:

> "You cannot make a machine think for you." This is a commonplace [expression] that is usually accepted without question. It will be the purpose of this paper to question it.... My contention is that machines can be constructed which will simulate the behavior of the human mind very closely. They will make mistakes at times, and at times they may make new and very interesting statements, and on the whole the output of them will be worth attention to the same extent as the output of a human mind.

In 1946, Von Neumann discussed the same question. "To what extent can human reasoning in the sciences be more effectively replaced by mechanisms?" He put the question and immediately made it more detailed and practical: "What phases of pure and applied mathematics can be furthered by use of large-scale automatic computing instruments?" His answer was very detailed and precise:

> Our present analytical methods seem unsuitable for the solution of the important problems arising in connection with nonlinear partial differential equations and, in fact, with virtually all types of nonlinear problems in pure

---

[92]These "pure" mathematicians also participated in the "flesh and blood" construction of the first automatic computers to suggest new mathematical and engineering ideas.

> mathematics. The truth of this statement is particularly striking in the field of fluid dynamics. Only the most elementary problems have been solved analytically in this field... The advance of analysis is, at this moment, stagnant along the entire front of nonlinear problems... Really efficient high-speed computing devices may... provide us with those heuristic hints which are needed in all parts of mathematics for genuine progress....

Then von Neumann stressed that the majority of these impregnable problems originated from physics. Usually some hints for solving them are given by experiments, but resources for making physical experiments are restricted and it is often difficult to interpret them. Using Frenkel's metaphor, we may say that it is not easy to draw a good "caricature" of a complex phenomenon, because there are too many superfluous details. On the contrary, all these details are naturally thrown away when we formulate the problem for a computer. We usually take a simplified mathematical model of the physical phenomenon that is free of unnecessary details (without destroying the main features), and make calculations under conditions imitating experiments. So to speak, we are modelling the real world on computers. This approach to solving problems of physics (and mathematics as well!) is called "computer experimentation" or "computer modelling."

The idea of the approach is clear. We are not just solving some equations describing something that is understood well. Our hope is to find (or to stumble across) a really novel phenomenon, something quite unexpected. In a way we are preparing everything for a play: we are familiar with the main characters and with their relations to each other, and we expect certain events to unfold. However, the play may develop by its own laws and we do not know where it will be in the end. The play is the play, not reality, but it somehow reflects reality and we could learn something from it.

Let us consider a physical example. Suppose that we are studying dislocations in the Frenkel–Kontorova model. Suppose that we stumble across a vague idea about movements of deformations ("thickenings" and "rarefactions") in the chain of atoms in a periodic external potential. Say we think they must move like "impulses" in the elastic string. Of course, we understand that the chain is a more complex system than the string, but we have to make a start! Now, we know the equations and it is not a big job for us to prepare a program for solving some initial-value problem (or, to adapt one of many existing standard programs). Then we start our game by suggesting different initial conditions. For instance, let two atoms be shifted from their equilibrium positions to the ones of their left neighbors. The computer will calculate the subsequent events and display them in the form of tables, pictures or movies, whatever you wish. Studying this output with a qualified eye, an experienced and thoughtful researcher might, just by chance, discover the birth of a soliton from a background of rather irregular motions — "the birth of a soliton from a foam." Or, if you are a very lucky person, you could see a soliton-antisoliton collision, or even a breather. It is not an easy job, but, having some idea of what you are looking for, you may discover something interesting.

Strangely enough, the rebirth of the soliton followed closely the above scenario. It was not so simple, because real computer experiments are not much simpler than the physical experiments. One needs good equipment: computers and other devices for handling information and presenting it in a comprehensible form; a team of professional mathematicians, programmers, and physicists to carry out the preparatory work in order to realize a serious computer experiment dealing with physics problems; a qualified staff of engineers and technicians to ensure reliability of the computer performance. It appears so intricate that you might ask: "Is it really a better way to look for a new physical construct than the good old method?"

Well, this question is not difficult to answer. First, the range of possible computer experiments is many times wider than that of physical ones. So, the probability of discovering new phenomena hidden deeply in the mathematical model is much higher. You can ask so many questions of the computer, sort out so many possibilities that, sooner or later if you are patient and lucky, you'll exclaim: "This is new territory!" It actually happened in this way with the soliton. It was discovered in computer experiments that were designed for studying very different things. The researchers had no idea of the soliton, but they immediately realized that something new had been found. "Look for India and you may find America!"

Second, and most remarkable, computer technology (hardware) and mathematical programming (software) are improving so fast that machines, which today are very compact and easy to handle, can do really difficult work. Without any exaggeration, we have begun to live our lives with computers. With a modern personal computer one can conduct experiments with the imagination of a Faraday or, at least, an Oersted.

I don't know if such experiments would interest Faraday or Oersted. Possibly, they would regard them as dull and abstract. But Euler or Gauss would be delighted, for sure! Both were exceptionally ingenious in calculations, and the idea of computer experimentation was not alien to them. As a matter of fact, the concept of computer experiments was quite clear to Euler:

> One may regard as rather paradoxical to ascribe any significance to observations in the branch of mathematics that is usually called pure mathematics. The common belief is that observations are restricted to physical objects accessible to our senses. Being confident that the numbers appertain to the domain of pure reason, we can hardly perceive any benefit from making experiments or quasi-experiments in studying the nature of numbers. Yet, most properties of numbers were discovered by observations. . . .

Following this credo, Euler implemented countless experiments with numbers (possibly, this titanic calculational work finally led him to blindness). Who knows how many discoveries he could have made with even a pocket calculator! Many modern scientists are taking advantage of the new opportunities provided by computers but "experimental mathematics" and "computer physics" are still not recognized as widely as they deserve.

Meantime, while some scientists are still trying to avoid computers or to use them only as convenient desk-tools, enthusiasts are collaborating with them in search of paths to new, unknown lands. One of these paths has led to the land of solitons. Strangely, this happened in studies of chaotic phenomena in physical systems. This is strange indeed, because solitons are lone creatures, fond of order, while chaos is an extreme manifestation of disorder. Then what related solitons to chaos? To answer this question we have to make a digression.

Many-faceted Chaos

> In vain to seek for a happy end
> In our times of chaos...
>
> Boris Pasternak

> Don't wake the sleeping thoughts!
> The chaos under them is stirring.
>
> Fyodor Tyutchev

We humans dislike disorder (having too much of it!) and the word "chaos" is almost abusive to us. I think that Boris Pasternak uses the word in this sense. The Russian poet Fyodor Tyutchev is speaking of a very different chaos. He also called it "ancient" and "native." Apparently, he had in mind something like the Chaos of the ancient Greeks — some primary state of the world from which everything was created. In modern terms, this must be the state of our Universe just after the "Big Bang." To understand the origin of our Universe — with all its galaxies, stars, planets — is one of the deepest unsolved problems of modern cosmology. It is probable that at some stages of the evolution of the universe, solitons helped to create order in the prime chaos. We mentioned an example when we discussed spiral sleeves of galaxies.

A different sort of chaotic motion is familiar from observing motions in water. The easiest way to create a disordered motion in water is to move some rigid body in it. Even for small velocity of the body, one can clearly see the vortices. For larger velocities, one can see a so called turbulent track, similar to those we like to watch behind a moving ship. For large enough velocity, the movements inside the track are completely chaotic; no order is left. Such movements were first studied by Kelvin, Boussinesq, Reynolds, and Rayleigh in the 19th century. The term "turbulence" was introduced by Kelvin, who derived it from the Latin word "turbulentus" (unquiet, disordered). Reynolds started the first experimental investigations of turbulent motions in 1893.

Turbulence is a very complex phenomenon. There exist several sorts of turbulence, having different types of disorder. For this reason all search for a simple and universal model of turbulence were unsuccessful for decades. The first successful models were invented in our, computer age, and the computer experiments were decisive for their exploration. This progress was very slow, because turbulence has its origin in nonlinearity of the equations, describing motions in fluids, and von Neumann's statement, quoted above, is fully applicable.

Having a simple computer (even a pocket calculator will do!), you can study the following elementary model of turbulent motion. The model has all the characteristic features of complex phenomena related to the generation of Chaos from Order, yet is very simple. There exist many similar models, but we consider one that requires only three rules of arithmetic. Imagine a clever, trained "flea." having a knowledge of the three arithmetic rules and arranging his/her jumps according to a certain law. Let the flea be sitting, at time $t_n = n\Delta t$ $(n = 0, 1, 2, \ldots)$, in a position x_n on the x-axis (we regard the flea as a point-like subject living in a one-dimensional world). Having some time Δt for calculations, the flea decides to jump, at the moment $t_{n+1} = (n+1)\Delta t$, to the position $x_{n+1} = b - x_n^2$, where b is a fixed number characterizing the "strength" of the flea (let us call this number the "flea-constant"). Suppose that the flea starts from some position in the interval $-2 < x < +2$, and we wish to know where it may be found after many jumps. In other words, what will be x_n for very large n?

Despite its simplicity this problem can hardly be solved without using a calculator. Still, some thought should be given to it before starting the numerical experiments. Let us try to find out what might happen to the flea when n is very large, i.e., $n \to \infty$. The simplest guess is of course that x_n has some limiting value x. Then this value has to satisfy the equation $x^2 + x = b$. The corresponding curve AA_0A_1 is presented in Figure 7.1. From it, you may find the limiting values x for each "flea-constant" b .

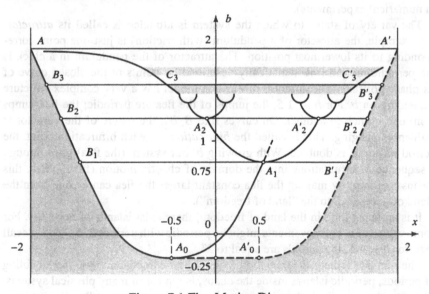

Figure 7.1 Flea Motion Diagram

Numerical experiments show that the life of a flea is not so dull. Only the fleas for which the points (x_0, b) are inside the curve $A_0B_1B_1'A_0'$ (the curve $A_0'A'$ is the mirror reflection of A_0A with respect to the b-axis) can approach the limiting curve AA_0A_1. Besides, they never reach the branch A_0A . If their parameters b and x_0 are such that (x_0, b) lies in the shaded region, they go to infinity. Those with (x_0, b) inside the curve $A_0B_1B_1'A_0'$ will end their life on the curve A_0A_1. So the fate of our clever fleas with $b < 0.75$ is predetermined: either to die at infinity or to be caught on the limiting curve A_0A_1.

Almost as sad is the fate of the fleas that start their jumps from the domain inside the curve $B_1B_2B_2'B_1'$. They finally come on the curve $A_1A_2A_2'$ jumping between two of its branches. The equation of this curve is easy to derive from the following condition: after any two jumps (for large enough n), the flea returns close to the initial position, i.e., $x_{n+2} \approx x_n$. As $x_{n+2} = b - (b - x_n^2)^2$, in the limit $x_{n+2} \approx x_n \approx x$ we find the equation $(b - x^2)^2 - (b - x) = 0$. The left-hand side of this equation is the product of $(x^2 + x - b)$ and $(x^2 - x + 1 - b)$. The first factor is zero on the curve AA_0A_1, while vanishing of the second one gives a new curve $A_2A_1A_2'$.

Up to now, the behavior of the flea has been more or less predictable. However, when the flea-constant b is increasing, the events become more and more complex. The curves to which the flea is attracted are bifurcating with smaller and smaller increment in b, and, for $b > 1.5$, the jumps are practically unpredictable, chaotic. In Figure 7.1, this is shown by the dotted strips. For instance, if the flea starts from a point on the line B_3B_3', it will be attracted to the line C_3C_3'. On this interval, the jumps are completely unpredictable (in fact, the movements of the flea in this domain are somewhat more complex, the reader may try to study them in numerical experiments).

The variety of states to which the system is attracted is called its *attractor*. For example, the attractor of a pendulum (with friction) is just one point corresponding to its lowermost position. The attractor of the pendulum in a clock is the periodic motion (mathematically described by points of the closed curve of its phase portrait). The attractor of our flea model has a very complex structure depending on b. For $b < 1.5$, the jumps of the flea are periodic; the flea jumps from one branch to another. On curves like B_2B_2', the nature of the attractor is suddenly changing. This is called the *bifurcation*. At each bifurcation point, the period of jumps is doubled. With growing b, our system (the flea) goes through a sequence of bifurcations into the domain of chaotic motion (for the flea, this means freedom; by making the flea constant larger the flea can escape from the "land of necessity" to the "land of freedom").

It is amusing that, in the land of freedom, there exist islands of necessity. For some values of b, the movements might be periodic with periods $3, 5, 7 \ldots$. Recall that, for $b < 1.5$, the periods are equal to 2^n.

The picture, namely the transition from periodic to chaotic motions, doubling of periods, periodic islands inside the chaos, is typical of many physical systems. The attractor on which the motion is chaotic is sometimes called the "strange attractor."

A thoughtful reader carrying out experiments on a computer might discover a beautiful regularity in the bifurcations. Let us denote the flea constants at the bifurcation points by b_n. Then $b_1 = -0.25$, $b_2 = 0.75$, etc. One may find that, for large n the differences of these numbers form a geometric progression. More exactly, for $n \rightarrow \infty$,

$$(b_{n+1} - b_n)/(b_n - b_{n-1}) \rightarrow 1/\delta, \quad \delta = 4.6692\ldots$$

This beautiful law was discovered by an American physicist, Mitchell Feigenbaum, in 1975. With his desktop computer, he studied a mathematical model, very similar to our flea-model, and stumbled on this law (of course, he had some idea of what to look for!). Then he carried out a thorough investigation of other, more complex systems, and finally he understood that a new law of nature had been discovered! It was not easy to convince other scientists, and for two years, science journals had rejected his paper. Fortunately, there are other means of communications, such as scientific conferences, and his discovery was finally recognized. Now, a quarter of a century later, the investigation of bifurcation and chaos phenomena is a highly respected, fashionable business. The new universal constant δ is now called *the Feigenbaum constant*, and his law is referred to as *the Feigenbaum similarity (scaling) law*.

Models, similar to our flea model are widely studied and applied to many interesting phenomena in physics and in other branches of science (we borrowed the flea model from a physics journal). This is, of course, only a first step to a real understanding of turbulence, but it appears that it is a correct one on the road to really understanding the nature of this extremely complex phenomenon.

> So small things help us to get an idea of great ones
> Hinting at possible ways to achieve them.
>
> Lucretius (ca 99–ca 55 BC)

Turbulent behavior often occurs even in very simple physical systems. This might look strange to physicists of the 19th century, who had regarded chaos as characteristic of systems of many particles. In fact, they thought that the number of the particles must be enormously large. The chaotic behavior of such systems looks quite natural if the particles interact. The first ideas of the heat phenomena (D. Bernoulli, M.V. Lomonosov) related them to chaotic movements of molecules (molecular chaos). These ideas were developed into a real physical theory through the efforts of Klausius, Maxwell, Boltzmann, Gibbs, and other scientists of the 19th century.

The theory of molecular chaos ascribes the chaotic behavior of molecules to frequent collisions between them. In these collisions, they exchange energies, and the values of their energies become close to some average energy, which is proportional to the temperature of the system. It is not difficult to conceive that, under normal conditions (e.g., in our room), the motion of the molecules of the air must be completely chaotic. We know that 1 liter of air contains approximately $n = N_A/22.4 \sim 3 \cdot 10^{22}$ of molecules (N_A is the Avogadro number). Their average velocity v is $\sqrt{3RT} \approx 200 \; km/h$ (R is the gas constant). The dimensional considerations allow us to estimate an average distance l, in which molecules do not suffer collisions. This distance depends on the size of the molecule and on the average number of molecules in a unit volume. It is approximately $l \approx 1/(d^2 n)$, i.e., $l \approx 10^{-5} cm$. It follows that in one second any molecule collides approximately $v/l \approx 10^{10}$ times!

Under these conditions, it would be meaningless to try to follow individual movements of any particular molecule. It is only possible to think of their *average* velocity and energy that define the gas temperature and pressure. But what will happen if the average number of molecules per unit of volume (determining the density of the gas) becomes much, much smaller? Or, what if we put the gas in a very cold environment? Clearly, with decreasing density or temperature, the collisions of the molecules will occur less and less often, and one may even expect that their motion will finally cease to be chaotic.

In such a case, average characteristics — pressure, temperature and density — may become useless and even meaningless, and so we may be forced to return from the average (*macroscopic*) description to a detailed (*microscopic*) picture. Up to which limit is the macroscopic description valid? This is the question to which the answer was unknown to Maxwell, Boltzmann and Gibbs. Even today, the question is too general and so has no universally applicable answer? Let us try to make it more precise.

What happens when the temperature decreases? At some low enough temperature, the gas will turn into a liquid and, at even lower temperatures, into a solid. Suppose that the solid is a crystal; let us analyze movements of its particles. To make things simpler, imagine a one-dimensional crystal, like a chain of small particles bound by springs. Now, the quintessence of the problem shows itself naked! The particles (molecules) are swinging near their equilibrium positions. But, are these movements ordered or disordered? In the previous chapters we analyzed, in some detail, rather regular motions in this system. Let us try to imagine what a disordered, irregular motion might look like. A moment's thought suggests that the amplitudes and maximal velocities of the oscillations must be very different and uncorrelated for different particles. The same has to be true for the phases. The average values of the amplitudes, velocities, and phases must be equal for all particles. We also expect that the average phase is zero. Is this possible for our model?

Most probably, your answer to this question will be negative. Recalling the theory of Chapter 4, it is easy to predict that the motion will be periodic. In fact, we have proved that the motions of the system of N bound particles can be presented as a sum of N harmonic modes. The motion in each mode is periodic, and the same must be true for their sum. More precisely, if only one mode of the frequency v_1 is initially excited then the period is $T_1 = 1/v_1$. If all modes are excited, the period is $T_N = 1/v_N$ where v_N is the minimal frequency. So, after some time our chain will return to the initial state, with the same positions and velocities of all particles. Then everything is repeated. Such behavior is certainly not chaotic.

A thoughtful reader might wonder about this point. In fact, in Chapter 5 we mentioned that, even in the "chain" of two particles, general motions might be aperiodic because the ratio of two frequencies is, in general, not a rational number. So a general motion with several excited modes must be aperiodic. The above reasoning tacitly assumes that the frequencies of the harmonic modes are proportional to the mode number $M = 1, 2, \ldots, N$. This is true for the low-frequency modes ($M \ll N$) in long chains of particles. Nonlinear dependence of the high frequencies on M gives rise to aperiodicity (we also know that it results in the dispersion of waves in the chain).

Well, all this is true, but the motion is still very regular; it is called quasiperiodic, meaning that the system regularly comes infinitely close to its initial state. Such a behavior can't be called chaotic. In the 19th century, many mathematicians believed that this was a general property of such physical systems as gases and solids. Some of them even made vigorous attacks on Boltzmann's hypothesis of molecular chaos. Boltzmann did not leave the arguments unanswered. Once he heatedly replied to his opponents: "Well, maybe the system returns to the initial state, but try to live long enough to see if this happens!" In fact, his belief in molecular chaos couldn't be mathematically validated in his time. But he followed his physical intuition, supported by the agreement of his theoretical predictions with experimental facts, so that he had reason to hope that mathematical problems could be somehow settled in future.

We are in this future, and the problems are nearly solved indeed! But the path to finding the solution has not been easy, and first steps once more led to solitons. The reason for this is that chaotic behavior can only be expected in nonlinear systems. Indeed, our chain model is linear, and so all motions of the atoms in the chain can be represented by superpositions of independent harmonic modes. There is no exchange of energy between different modes and, therefore, there is no mechanism that could make mean values of the energies of individual harmonics equal.

Now, in the real physical crystal, there exist nonlinear forces between atoms, because the linear Hooke's law is only a first approximation (the effective nonlinearity in gases is provided by collisions of molecules). Then, it is reasonable to expect that the harmonic pattern of motion is only a first approximation. Even very small nonlinear forces will provide exchange of energy between the modes, and this will result in establishing a chaotic (thermodynamic) state. At least, one would expect this for large enough N. Probably, when Enrico Fermi began one of his last works in science, he had in mind something like this. A debatable point is, of course, "large enough" N. Is $N = 100$ large enough? Maybe, it is better to think of numbers like $N = 1000000$, or even $N = 10^{23}$? I think Fermi guessed that there must be a sort of bargain between N and the strength of the nonlinear forces. The larger the force, the lesser the N at which chaos may be established. Anyway, his first question to the computer was about approaching a chaotic (thermodynamic) state in a chain of particles bound by nonlinear springs. The result of his computer experiment was very puzzling.

Enrico Fermi is Astonished by the Computer

Almost everyone using computers has experienced instances
where computational results have sparked new insights.

Norman J. Zabusky

Enrico Fermi was one of the greatest physicists of the 20th century, equally great
in theory and experiment. He will be remembered forever for his contribution to
exploring nuclear power, to the physics of elementary particles and for his many
other contributions to various branches of physics. Modern physicists mention his
name everyday — Fermi statistics, fermion, Fermi surface, Fermi liquid, etc. It is
less widely known that he was interested for many years in nonlinear problems
and in some problems related to chaos. In fact, one of his first scientific works
dealt with some problems of statistical mechanics. Not long before his untimely
death, he turned to problems of turbulence and carried out several works in hydrodynamics (one of them in collaboration with von Neumann). Possibly, not without the influence of von Neumann, he started his exploration of *terra incognita* in
nonlinear physics through computer experiments.

We describe his work, which had been done at Los Alamos in the laboratories
created for the Nuclear Project. Nowadays, Los Alamos is one of the main centers
for nonlinear studies, and nonlinear physics is among the main topics of research
at the Center for Nonlinear Science. It was not by chance that Feigenbaum made
his discovery while working there. Even the most original and independent minds
need some "atmosphere" for their work. It is not easy to formulate in exact terms
what this "atmosphere" is. Even more difficult is to create a proper atmosphere.
Anyway, it is clear that Los Alamos had quite the right atmosphere for Feigenbaum, and its creation can be traced to the old work of von Neumann and Fermi.

Let us turn to the problem that Fermi suggested to the computer. In collaboration with mathematicians Stanislav Ulam and John Pasta, he planned to perform
a variety of computer experiments with nonlinear problems. The first was about
the origin of chaos in the nonlinear chains of atoms. Later Stan Ulam recollected:

> The original objective had been to see at what rate the energy of the
> string, initially put into a single sine wave (the note was struck as one tone),
> would gradually develop higher tones with the harmonics, and how the shape
> would finally become "a mess" both in the form of the string and in the
> way the energy was distributed among higher modes. Nothing of the sort
> happened. To our surprise the string started playing a game of musical chairs,
> only among several low notes, and perhaps even more amazingly, after what
> would have been several hundred ordinary up and down vibrations, it came
> back almost exactly to its original sinusoidal shape.
>
> I know that Fermi considered this to be, as he said, "a minor discovery".
> And when he was invited a year later to give the Gibbs Lecture (a great
> honorary event at the American Mathematical Society meeting), he intended
> to talk about this. He became ill before the meeting, and his lecture never

took place. But the account of this work, with Fermi, Pasta and myself as authors, was published as a Los Alamos report.

Let me explain this work in more detail. Fermi, Pasta and Ulam (FPU, for short) asked the computer to derive vibrations in the system of 32 massive particles bound by springs. They accounted for nonlinear forces by assuming that stretching the spring by Δl generates the force $k\Delta l + \alpha(\Delta l)^2$. The nonlinear correction to Hooke's law, $\alpha(\Delta l)^2$, was assumed to be small as compared to the linear force, $k\Delta l$. Thus, the computer had to solve equations like (5.8), but with additional nonlinear terms in the right-hand sides:

$$\alpha[(y_{l+1} - y_l)^2 - (y_l - y_{l-1})^2].$$

As these corrections are small, one may take as a first approximation the solution of the equations (5.8) obtained for $\alpha = 0$. For small values of α, the harmonic modes still approximate the exact nonlinear equations but become interacting. This means that, in the time intervals, which are small in comparison with the periods of the modes, they slowly exchange energies. We may consider the modes as analogues of molecules in a gas and the nonlinear forces as analogues of collisions (like collisions, these forces provide the energy exchange). Then we expect, by this analogy, that the energy will eventually be uniformly distributed between all modes and the system will completely forget its initial state. This means establishing a chaotic motion. Fermi, Pasta and Ulam expected something like this. Let me briefly describe what they actually saw.

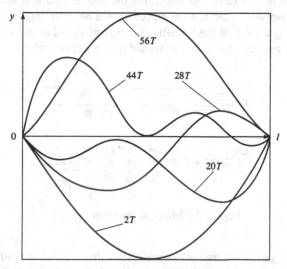

Figure 7.2 Positions of the particles in the Fermi–Pasta–Ulam
chain for different times

Consider an initial state, in which only the first mode is excited (let us denote its period by T). Then the energy slowly goes into other modes (see Figure 7.2).

For $t \approx 20T$, mainly the third mode is excited. Then the second mode appears on the stage (for $t \approx 28T$). For $t \approx 44T$, the energy is mainly concentrated in the third mode and, for $t \approx 56T$, the first mode returns. Sometimes higher modes are slightly excited but the maximal energy of the fourth mode is less than a half of the first one, while the fifth mode may acquire less than a sixth part of the total energy. Other modes are practically unexcited.

The return to the original state is not accidental. Computations with other initial states, with different values of α, give similar results. Even adding a cubic nonlinear force did not change the main feature — returning to the original state. Modes are like soloist playing in turn. At a given moment there is one soloist, other instruments softly accompanying the solo performer. When the first finally returns, the whole piece is repeated. The time of return (in the described case it is $\approx 56T$) depends on N, as well as on the form and the strength of the nonlinearity, but it occurs for a wide range of these parameters.

One may visualize the result by writing a simple musical piece. In Figure 7.3 we represent modes by notes. To each mode corresponds a note. To the first one — lower "do," to the second — "do" an octave higher, etc. Here the energy of the mode is depicted by the duration. This is inconvenient if you try to play this piece, but otherwise it is a musical piece.

So, instead of expected cacophony (when all modes are equally loud), we have a primitive but still musical piece. This is quite unexpected for a nonlinear system. To any initial excitation it responds by a "chorale." If we start with a second mode, there will be another "chorale," etc. So this nonlinear string is like an improvisor. It would not be difficult even to transform the modes into real sounds, thereby making these chorales audible. Real pieces would be more interesting than what is depicted in Figure 7.3, as the transitions from one chord to another are smooth. In addition, the excited higher modes would give a richer timbre.

Figure 7.3 Musical computer piece

Several other physicists and mathematicians also found this result fascinating and began to try to understand it. Martin Kruskal and Norman Zabusky heard of the results from the authors and immediately started similar computer experiments. First, they repeated FPU experiments, then considered movements of continuous nonlinear strings (the continuum limit for the FPU chain). After many attempts, they came to a striking conclusion: for small amplitudes, vibrations of the continuous FPU string [93] are best described by the Korteweg–de Vries equation!

Russell's Soliton Returns

> Long years you waited
> Until your hour came...

Valerii Bryussov (1873–1924)

Recall that Korteweg and de Vries discovered their equation in attempting to find an exact mathematical description of the small amplitude soliton. Zabusky and Kruskal have shown that the same equation describes quite different physical phenomena. This is not accidental. In fact, the KdV equation describes a variety of nonlinear waves, and is suitable for small amplitude waves in materials with weak dispersion.

If both nonlinearity and dispersion are negligibly small, the KdV equation becomes linear. It describes waves moving in one direction, so that the shape of the wave is $y(t, x) = f(x - v_0 t)$, where f is an arbitrary function of one variable. With positive v_0, this describes only right moving waves (the left moving waves are described by the same equation with $v_0 < 0$).

For shallow water waves, $v_0 = \sqrt{gh}$, where h is the depth. Recall that this approximation is valid if the minimal wavelength λ, of the harmonic waves in the Fourier expansion of f, is much larger than the depth h. We may forget about Fourier expansion by assuming that the wave is approximately sinusoidal.

To account for small dispersion, we may write the simplest possible correction to v_0 in the form suggested by (5.17) and (5.21)

$$v = v_0(1 - \alpha \frac{h^2}{\lambda^2}).$$

Here, α is some number, and $v_0 = \sqrt{gh}$. For shallow water waves, $\alpha = (2/3)\pi^2$. Note that this general expression is applicable to many other waves.

In Chapter 5, we compared shallow water waves with those in the chain of atoms bound by springs, (5.21) and (5.17). There we saw that dependence of the wave velocities on wavelength is of the same form in both systems. Just replace h by $a/2$, where a is the distance between the atoms, and you'll obtain the dispersion law for long waves in the chain. This fact is amazing but, once you know it, you can guess that the dispersion law might be similar in other physical systems. So it is equally astonishing that nobody guessed this for decades! The only explanation I can give is that waves in crystals and waves in water were studied by scientists belonging to distinctly separate scientific communities (so to speak,

---

[93] Of course, any computer calculation uses some kind of a discrete approximation to the differential equations but it can be made as exact as you wish.

"liquid-ists" and "solid-ists"). Anyway, a more or less clear idea of a universal dispersion law was born not long ago, in the soliton age.

To understand KdV waves, consider a simple nonlinearity. We know that the velocity of linear waves in dispersive media depends only on their wavelength, but it doesn't depend on their amplitude. On the contrary, the velocity of nonlinear waves must depend on their amplitude. The simplest possible dependence is linear, when the velocity is proportional to the amplitude. This very dependence is realized for the KdV waves. Being the simplest one, it naturally is found in many other systems. Zabusky and Kruskal showed that this dependence is valid for lattices of atoms. Even earlier, in 1958, the Russian physicist R.Z. Sagdeev noticed an analogy between shallow water waves and some waves in plasma. He also discovered solitary waves in plasma. At that time, many physicists were studying plasma, having in mind controllable thermonuclear reactions, and his discovery was not unnoticed. It soon became clear that these waves are also described by the KdV equation.

This was really fortunate for KdV. It was recalled from oblivion and soon became famous among physicists and mathematicians. This happened after the striking discovery of Zabusky and Kruskal who proved that the KdV solitons are not changed in collisions, like rigid bodies. No less amazing was the discovery made in 1967 by Gardner, Greene, Kruskal, and Miura who found the general solution to the KdV equation which triggered a wide interest in soliton theory.

The KdV equation does not look particularly complicated, and we write it here for interested readers. The shape of the wave $y(t, x)$, at any moment t, satisfies the equation

$$\dot{y} + v_0 \left(y + \frac{3}{4h} y^2 + \frac{h^2}{6} y'' \right)' = 0.$$

Here the dot denotes the derivative in t for fixed x, and the prime denotes the derivative in x for fixed t. If we draw the profile of a wave given by some solution of the KdV equation, the obtained picture will move along the x-axis. In general, the profile will also deform. If we know the shape of the wave, $y(t_0, x)$, at any given moment $t = t_0$, the equation tells us how the wave will move and deform. Indeed, we can find all x-derivatives of $y(t_0, x)$ and then find $\dot{y}$ by using the equation. Then, as

$$y(t_0 + \Delta t, x) \approx y(t_0, x) + \dot{y}(t_0, x)\Delta t,$$

one can draw the profile at the moment $t + \Delta t$ (for small Δt, this is simply the definition of the partial time-derivative).

To solve the KdV equations means finding $y(t, x)$ for any given initial shape $y(0, x)$. For nonlinear equations, this is a very difficult problem because there is no superposition principle (a sum of solutions is not a solution, unlike in linear equations). So the discovery of methods providing *exact* solution of this problem is one of the most remarkable achievements of mathematics in this Century. This discovery became possible as a result of a fruitful collaboration between physicists, mathematicians, and computers. Today we know several classes of exactly solvable nonlinear evolution equations, some of them are mentioned in the Appendix.

It is easy to understand that the KdV equation describes waves running in one direction on the x-axis. First, notice that dispersion effects depend on the y''–term in the equation, while nonlinearity is given by the y^2 term. Neglecting both of them will give a very simple equation $\dot{y} + v_0 y'$, which we discussed in Chapter 5. There we found the most general solution of it: $y(t, x) = f(x - v_0 t)$. To find the dependence of y on x at any time, one has simply to plot the curve $y = f(x)$ and move it with velocity v_0 in the positive direction of the x axis. To obtain a wave moving in the negative direction, one has to change the sign of the velocity. But this means that we have to change the equation. We thus conclude that waves travelling in different directions satisfy different equations! Doesn't it look somewhat strange?

You must remember that the D'Alembert equation does not depend on the sign of v_0, and so describes waves running in both directions. The same is true for the equation describing dislocations. Moreover, everyone knows that the waves created by a stone thrown into a tiny brook run both upstairs and downstairs (or, to the left and to the right) and they are obviously identical. Then, why does the KdV equation depend on this sign?

The answer to this apparently difficult question is in fact very simple. If we are not interested in what is happening at the initial moment, we may study left moving and right moving waves independently. For small enough dispersion and nonlinear terms, left and right moving waves will be well separated before these terms have any influence on them. So, we can study one of them, completely forgetting the other. Note that this simplification is only possible when dispersion and nonlinearity are small. For dislocations, when dispersion and nonlinearity effects are large, this independent treatment of left and right waves is impossible. The idea of disentangling left and right movers was crucial for Korteweg and de Vries in their treatment of Russell's soliton. Simplifications provided by this simple (but not obvious!) idea 70 years later allowed for the construction of a complete mathematical theory of solitons and opened ways to today's applications.

From the physicist's point of view, the most important feature of the KdV equation is that the role of dispersion and nonlinearity in forming solitons is quite obvious, and one may clearly see how balancing these effects results in forming a stable solitary wave. Let us look at the KdV equation from a physics viewpoint. If there were neither dispersion nor nonlinearity, impulses of any shape might travel over the water surface. But these are not solitons. The smallest nonlinearity or dispersion, which exists in any physical medium, will eventually change its shape beyond recognition. Let us first look at the role of nonlinearity.

Remark in passing that neglecting the dispersion term will make the KdV equation easy to solve. The method for solving the so-truncated KdV equation was known even to Lagrange, but first applications can be traced back to the famous German mathematician Georg Friedrich Bernhard Riemann (1826–1866).

Let us imagine a small hump on the surface of the water (sketched in Figure 4, curve 1).

The subsequent evolution of this hump is defined by dependence of the velocity of each point on its surface on its height (which is $y(t, x)$). For the KdV waves,

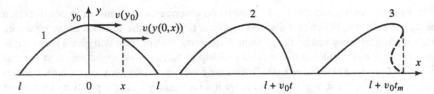

Figure 7. 4 Evolution of a hump on the surface of the water

this dependence is very simple:

$$v(y) = v_0 \left[1 + \frac{3}{2} \frac{y}{h} \right].$$

The fastest is the top of the hump, having velocity

$$v(y_0) = v_0 \left[1 + \frac{3}{2} \frac{y_0}{h} \right].$$

The velocity of the front of the hump, where $y = 0$, is of course v_0. It follows that the front is becoming steeper (curve 2) and, eventually, its top will fall down (an "overturn"). This "overturning" cannot be described in such simple terms, but all of us have seen this extreme manifestation of nonlinearity while watching the surf at a seaside or, on a smaller scale, on a river sandbank.

> And rocks are more and more sheer,
> And waves are steeper and sterner...
>
> Heinrich Heine (1797–1856)

It is not so simple, and even dangerous, to watch another phenomenon having the same origin. When a tidal sea wave enters the mouth of a river, a big wave having the shape of a very steep step may arise (see Figure 7.5).

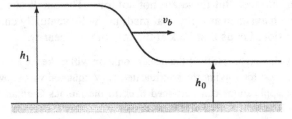

Figure 7.5 Sketch of a tidal wave

This wave, sometimes called a "bore" is an example of a shock wave. A more familiar example of the shock wave is a loud clap caused by a speeding up jet fighter. This signals that it is exceeding the speed of sound or, in other words, it is getting over the "sound barrier." You may prefer a more idyllic example — the crack of a cowboy whip, which is of the same origin.

Shock waves were theoretically predicted by Riemann in 1860 but he was uncertain about the possibility of observing them. Meanwhile, Russell noticed that the sound of gunfire from a distant gun reached his ear before the command "Fire!" He had correctly guessed that the speed of sound of gunfire exceeds the normal one. In fact, he observed an important general property of shock waves that makes them easy to detect.

The speed of the tidal wave shown in Figure 7.5 is $v_B = \sqrt{gh_1(h_0 + h_1)/2h_0}$. As $h_1 > h_0$ we have $v_B > v_1 = \sqrt{gh_1} > v_0 = \sqrt{gh_0}$, and so the tidal wave travels faster than any other small wave on the surface. Note that the shock wave in air (or, explosive wave) is a more complex physical phenomenon than the tidal wave. On the wavefront of the explosive wave, there is a sharp jump in pressure, temperature, and density. The maximal value of the density cannot exceed a certain limit, but the pressure and temperature may be very high. For example, if the maximal pressure in the shock wave is equal to 100 normal atmospheric, the temperature may be as large as 3500 degrees centigrade. At this temperature, violent internal vibrations in the molecules are excited and some oxygen molecules start to decay into atoms. This gives rise to many chemical reactions. For even higher temperatures, the shock wave is brightly shining. An apocalyptic view of the shock wave ("brighter than a thousand suns") is, unfortunately, too well-known to us from numerous descriptions of the nuclear explosion. The temperature in that shock wave is above 10 000 degrees and the pressure is about 1000 normal atmospheric.

A tremendous shock wave had been created on July 30 of 1908 by the famous[94] Siberian ("Toongoossky") meteor. Several seconds before its fall, occasional spectators had observed a blinding fire ball, followed by the deafening sound of an explosion. This "big bang" had been heard for a range of 1000 kilometers, and the shock wave of this explosion had been registered even in England! Modern estimates tell us that such a shock wave can be produced by a several megaton nuclear explosion. But, most probably, that wave was not from the explosion. Its nature was similar to that of the jet fighter shock wave, i.e., "overcoming the sound barrier." A massive body, possibly a small comet or a piece of a comet, had entered the Earth's atmosphere with a super sound speed and created a terrifying shock wave (it was a real shock for the spectators!).

The reason for the bolide (fire ball) shining is also the shock wave. A shining shock wave is similarly produced by satellites slowing down in the atmosphere. This, by the way, is very useful: a big portion of a satellite's kinetic energy is spent on forming the shock wave. So it works like brakes or a parachute. You might have heard fairytale-like stories about shock waves from war veterans. I read one in a popular science brochure by A. S. Kompaneets, a well-known specialist in the theory of shock waves. He narrates the story of the miraculous salvation of a Russian pilot during the Second World War. His fighter plane was shot down and his parachute failed to open. His death seemed inevitable but, a moment before

---

[94]I remember seeing posters of several lectures in Moscow of the fifties: "The Mystery of the Toongoossky Meteor." They were at the time as popular as the lectures: "Is there Life on Mars?"

crashing to the ground, a bomb exploded just under him. The shock wave of this explosion saved his life.

Unfortunately, the majority of scientific studies of shock waves are related to developing and using modern weapons. In spite of my great respect for this interesting science and its creators, I would prefer to avoid further discussions of this topic. Let us return to our peaceful solitons.

Consider in more detail the bargain between dispersion and nonlinearity resulting in the soliton formation. The initial hump shown in Figure 7.4 (the first curve) may be represented as a sum of harmonics. The wavelength of the first harmonic is about two widths of the hump. The wavelengths of the higher harmonics, which are needed to form a hump, are larger. Due to dispersion, the second harmonic moves faster than the first one, the third faster than the second one, etc. The result is that the slope of the hump's front is softened by dispersion. Remembering that nonlinearity is making the slope steeper, it is easy to imagine that, for some shape of the hump, there exists a velocity for which the hump can move undeformed. In other words, the hump is "tuning" its shape and velocity so as to reach a stable motion, in which there is a complete balance between the deforming effects of dispersion and nonlinearity. Roughly speaking, this is how the soliton is formed. If the initial hump is too big to form just one soliton, it will break down into several pieces. If it is too small, it will diffuse because the nonlinear effects cannot balance the dispersion.

We can now write an exact equation for this balance. First, let us look at the exact solution of the KdV equation describing the soliton:

$$y(t, x) = y_0/\cosh^2\left(\frac{x - vt}{l}\right). \tag{7.1}$$

Here $v = v_0[1 + (y_0/2h)]$, and l is defined by the equation

$$S \equiv \frac{3}{4}\frac{y_0 l^2}{h^3} = 1. \tag{7.2}$$

This is the mathematical condition expressing the balance between the dispersion and nonlinearity effects in the soliton. Although the parameter S was known for decades, its role in soliton theory became clear fairly recently.

To understand the origin of this simple condition, recall that nonlinearity is adding a term of the order of $v_0 y_0/h$ to the speed of the top, while dispersion is slowing it by $v_0 h^2/l^2$ (recall that the velocity of the hump's top is mainly determined by that of the first harmonics having the wavelength $\approx 4l$). These would-be changes compensate each other if the condition (7.2) is satisfied.

If S is much larger than 1, nonlinearity effects will prevail (if the hump is smooth enough). They will strongly deform the hump and it will eventually break into several pieces that will probably give rise to several solitons. If $S < 1$, dispersion prevails, and the hump will gradually diffuse. For $S \approx 1$ the hump has a soliton-like shape and, if its velocity is close to the soliton velocity, it will slightly deform and gradually become the real soliton, described by (7.1).

Recalling that $\cosh x$ is a positive function, having the minimal value for $x = 0$, $\cosh 0 = 1$, and that it is increasing very fast with $|x|$, you may easily draw the shape of the KdV soliton (or, just ask your computer to draw the function of x given by (7.1) for some fixed time). It is clear that the position of the soliton's top (its center) is at $x = vt$, and that we may call its width $2l$. The outer part of the soliton (where $|x| > l$) is usually called its "tail." It is natural to call its central part the "head." About 90% of the water, forced by the soliton over the water surface, is in its head.

In the first part of this book, we mentioned Russell's formula for soliton speed, $v = \sqrt{g(y_0 + h)}$. At first sight, it differs from the exact one written after (7.1). In fact, there is no contradiction, if we take into account that the amplitude y_0 is small in comparison to the depth h. So,

$$\sqrt{g(y_0 + h)} = \sqrt{gh(1 + y_0/h)} \approx \sqrt{gh}(1 + y_0/2h).$$

Now, we may return to the Zabusky–Kruskal numerical experiments with the KdV equation. When they set up the experiments, they had a good knowledge of the soliton solution and of the nonlinear periodic waves. It was less clear how the solitons could be generated. What would happen with two colliding solitons was quite unknown to them. They probably believed that the solitons would be destroyed or, under special conditions, a new soliton would be created by emitting a redundant energy in waves.

Their approach to the problem was similar to that of Fermi, Pasta and Ulam. Choosing some initial shape of the surface, $y(0, x)$, they computed its evolution in time (as a matter of fact, they studied waves in plasma satisfying the KdV equation). To make the problem accessible to the computer, Zabusky and Kruskal had to restrict the x axis to a finite interval $0 \leq l \leq L$. To do so they imposed on $y(t, x)$ certain boundary conditions, accounting for the reflections of waves from the edges of the interval. These conditions are defined by physics but, to simplify the problem, they applied the standard trick of imposing periodic boundary conditions. [95]

Consider the evolution of a simple harmonic wave $y(0, x) = \cos(2\pi x/L)$. This initial shape of the wave is shown in Figure 7.6 by the hatched-dotted line. At some moment T, a characteristic step is formed (the hatched curve), which later at time $2.5T$, gives rise to a sequence of solitons (the solid curve). The solitons are enumerated in descending order, the first having the largest amplitude. They all move to the right, and the one crossing the right border immediately appears on the left border (recall that they are moving on a circle, i.e., the points $x = 0$ and $x = L$ should be identified, like the "ends" of the equator on the map of the Earth). The first soliton is the fastest one, and so it is running down other solitons and eventually colliding with them. This is roughly shown in the diagram Figure

---

[95]This means that $y(t, 0) = y(t, L)$ and $y'(t, 0) = y'(t, L)$, which is equivalent to considering waves on a circle.

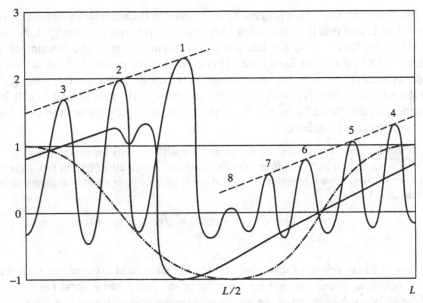

Figure 7.6 Evolution of the harmonic wave

7.7. (Figures 7.6 and 7.7 are reproduced from the original paper by N. J. Zabusky and M. Kruskal.)

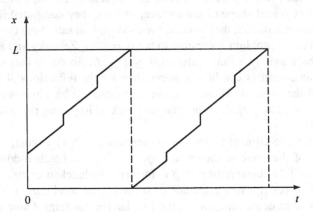

Figure 7.7 Path diagram for the first soliton

The small steps correspond to the collisions. One may clearly see that the soliton accelerates at the moment of the collision (look at the sharp jumps). After each collision it assumes the same velocity. We discussed this process in more detail in Chapter 2, where we used the tennis ball analogy for the soliton.

All these results have been really amazing. An especially deep impression was produced by the movie, in which one could see how the solitons are created, what happens to them in collisions, etc. This first movie, which opened the intimate life of solitons to the public, was in black and white and not sound tracked. Today, there exist many colored and sound tracked movies on solitons and other nonlinear phenomena. A film festival might be even arranged that would give more information about solitons than several books like this one. I foresee that in a not too distant future, computer-made video films will possibly compete with popular science books. Yet, I hope that the art of storytelling will not completely disappear, and thus I proceed with my story.

In fact, we are approaching the end. After the fundamental work of 1965–1967, the mathematical theory of solitons was developed by the combined efforts of scientists in USA, USSR (now, FSU), UK, Japan and other countries. This really international enterprise has laid the foundation of a new branch of mathematical physics, with many new beautiful ideas and results.

Unfortunately, in this book I can barely touch upon these matters because the needed mathematical equipment is too complex. To avoid using advanced mathematical theory, I will mainly describe physics embodiments of the soliton idea. It is also not an easy task. These embodiments are so numerous and diverse that any attempt at completeness would be hopeless. As a compromise, I will give only an impression of physical applications of the solitons, which have been discovered in the last 30 years, but I will have to leave out many interesting things.

I will mainly attempt to trace the development of 19th century ideas, described in Chapter 3. In this way, we shall meet some new applications, but this will be a cursory inspection; do not expect too much of it. It will be like a quick bus tour in Moscow: "On the right is the Bolshoi Theater building, constructed in 1824, designed by the architects Bove and Mikhailov; on the left, we see the Maly Theater building and Ostrovsky's statue, made by the sculptor Andreev; ahead is the hotel Metropol with the famous ceramic panel designed by Vrubel. . . ."

But let us first finish with the FPU computer experiment. It is worth mentioning that Fermi, Pasta, and Ulam were very fortunate. Could they have made computations on a much larger time scale, they would have found that the observed recurrence to the initial state was not exact! For very large time intervals, this recurrence would be less and less visible, and finally the chaotic state they were looking for would be established.

The system considered by Zabusky and Kruskali is quite different in that the recurrence to the initial state is exact, at least in principle. We say in principle because, in any real computation, there are small errors that become important in computing the evolution on large time intervals. In fact, having small enough computational errors, one will really see that, after many collisions, the solitons disappear and the initial sinusoidal wave returns!

The systems in which initial states, sooner or later, return [96] belong to a class of *integrable* systems. For instance, all physical systems with true solitons are

---

[96]We assume that the coordinates describing the system belong to finite intervals. The free particles in space never return to the initial positions, but the system of these particles is also integrable.

integrable. You may easily guess that all linear systems are integrable and that the simplest example of an integrable nonlinear system is a pendulum. This statement tacitly assumes that there is no friction, which makes any physical system non-integrable. However, if there is little friction, the pendulum is an almost integrable system, meaning that friction can be treated as a small perturbation.

The variety of almost integrable systems is much wider than the set of the exactly integrable ones, and this makes studying integrable systems interesting for applications (in general, nonlinear systems are not integrable in any sense). The Fermi–Pasta–Ulam chain is almost integrable, but there also exist exactly integrable chains with particles coupled by nonlinear forces. In 1967, a Japanese physicist, M. Toda, discovered an exactly integrable, essentially nonlinear chain. If the force, acting on the nth atom, has the exponential form,

$$a\left[e^{-b(x_n - x_{n-1})} - e^{-b(x_{n-1} - x_n)}\right],$$

the chain is integrable and admits solitons similar to the KdV (Russell) solitons. The Frenkel–Kontorova model is integrable in the continuum limit, and so it is almost integrable. Many other examples might be added, but it is time to begin our brief soliton tour!

Ocean Solitons: Tsunami, the Tenth Wave

> And waves, invariably angry,
> Are running fast against my shore.
>
> Aphanasii Fet (1829–1892)

On December 28 of 1908, a terrible earthquake killed 82,000 people in southern Italy. A monstrous tidal wave, triggered by this earthquake, was described by M. Gorky: "... An immensely high wave got up to the sky, shielded half of the sky. Then, swinging its white crest, it bent, broke in two, collapsed on the beach and, with all its dreadful weight, ... washed away all the coast." Gorky described here the *tsunami*. The Japanese word "tsunami" means "a big wave in a harbor." Usually, tsunami is produced by a large enough sea soliton. It is quite harmless in the open sea, but becomes very dangerous when coming to the shore. There may exist tsunami not directly related to solitons but experts agree that the majority of registered tsunamis were produced by solitons.

Usually solitons in the sea are produced by earthquakes having epicenters at the sea bottom. The length of such a soliton may be as large as 500 kilometers but, even for the 10 kilometer long soliton, the ocean may be regarded as shallow. We thus can use our knowledge of shallow water waves. Normally sea solitons are not very high, less than 10 meters or so. Such a long and smooth wave is not easy to recognize. An average velocity of the soliton is $v = \sqrt{gh}$. For $h \approx 1\ km$ we thus find $v \approx 100\ m/sec$. The velocity varies with depth, usually from 50 to 700 kilometers per hour. Coming to the coast, the soliton slows down and becomes shorter and higher.

To see this, recall (7.2) and take into account that the soliton energy does not change with its deformations. The energy is proportional to $y_0^2 l$; let us denote this quantity by L^3, where the new parameter L has the dimension of length. Using (7.2) with $S \approx 1$, we find that $y_0 \approx L^2/h$ and $l \approx h^2/L$. Ocean solitons normally have about several dozen L meters and, as $h \sim L$, the tidal wave could become rather high. These simple estimates are only valid for $L < h$. When $L \geq h$, nonlinear effects are not small and lead to collapsing of the wave, as described by Gorky.

Tsunami most often "visit" Japan and Chile. One of the strongest tsunami struck Japan in 1896 and killed 30,000 people. The tidal wave was about 30 meters high. Among the newspapers readers' accounts of the chilling details of this tsunami were possibly Korteweg and de Vries as well as several scientists familiar with their paper, which had been published a few months earlier. However, nobody, at that time, could see a relation between these two events. These dangerous tidal waves escaped scientists' scrutiny for quite some years; the soliton theory of the tsunami phenomenon was proposed three quarters of a century later.

Modern theories of tsunami deal with three main problems. The first and most difficult is to predict the soliton formation by earthquakes (on the sea bottom, near a sea coast, etc.), by volcanic eruption, landslides on sea coasts, nuclear weapon explosions, etc. All these catastrophes can generate potentially dangerous solitons. For example, the eruption of Krakatoa in 1883 produced a 45 meter high wave on Java and Sumatra islands killing 36,000 people. Experts believe that the energy of the eruption was comparable to that of a few hundred Hiroshima explosions. The second problem concerns propagation of the soliton. It is the easiest one, because we have a good mathematical theory of the soliton as well as many tools, including scientific satellites, for detecting big waves in the ocean. The third problem is to predict the behavior of the soliton while approaching the coast. It is almost as difficult as the first one, because this behavior depends on many factors that vary from place to place. For instance, the same wave will behave differently when coming to a smooth sandy beach, to a harbor, or to a river mouth.

Thus, the diversity of causes and manifestations of tsunamis makes their scientific analysis rather complex. The analysis of tsunanis involves many branches of modern science — physics, mathematics, geology, and oceanology. To predict tsunami, we need a number of modern technical tools, but even with satellites, supercomputers and global systems of telecommunications, we are still incapable of fully solving this problem. Many laboratory and computer experiments are being carried out by scientific teams studying tsunami, and new theoretical approaches are being proposed but, as yet, there is no absolute guarantee against disaster from this dangerous soliton.

There exist "small tsunamis," created by moving vessels. These became known in the beginning of the 20th century to captains of fast warships. From time to time, they observed that a backwave, created by the ship, moving at a speed close to $\sqrt{gh}$, separates from the ship and starts moving by itself. It is not difficult to guess that this backwave is Russell's soliton. The only difference is that, in a shallow canal, this wave can be created by slow moving vessels (like a barge).[97] The cause for breaking of the backwave off the ship may be either the ship's sudden slowing down or its coming close to a bank (if the depth suddenly changes). Then the wave may start to walk by itself, and it may be dangerous for those who happen to be around.

In 1912, A.N. Krylov investigated such an accident, in which a child had been drowned by a backwave soliton from a cruiser, on a beach near St. Petersburg. He described this accident in his very interesting memoirs. Unfortunately, he was not familiar with the soliton concept and only gave practical advice to captains.

Many different solitons travel in the ocean. Being infinitely diverse and complex, the ocean provides a natural living environment for them. Some solitons have been considerably studied but there are many others, surely enough for much exploration, not yet discovered. Waves and solitons can exist in places other than the surface of water. Deep seas have a rather inhomogeneous structure — there are layers of different temperature, density and saltiness. Quite often borders between different layers are fairly sharp; in this case, they are called "dividing surfaces." These dividing surfaces have many features in common with the sea surface, which is the dividing surface for water and air. Thus, waves and solitons can also travel over the dividing surfaces. Like tsunamis and backwave solitons, they may be very dangerous. One may entertain the possibility of explaining some mysterious affairs with submarines by their collisions with deep water solitons, but I have never heard of any serious discussions of such events; this information is probably classified.

> Oh my sea wave,
> You are willful.
> Both in peace and joyful game,
> You are full of lovely life.
>
> Fyodor Tyutchev

There is one more soliton on the water surface that is known to men from time out of mind, but which has escaped scientific scrutiny until very recently. Everybody is familiar with *groups* (*packets*) of waves, created by the wind on a deep water surface (Figure 7.8). You have possibly observed waves coming to the beach from an apparently quiet sea and wondered where they are coming from.

The phenomenon is so obvious and amazing that it would be very strange if scientists did not try to explain it. Of course, they tried and, in particular, at-

---

[97]Russell ascribed a first observation of the soliton, created by a moving barge and the resulting diminishing of water resistance to a "William Houston, Esq." This gentleman, however, recognized only a "commercial significance of this fact for a canal company for which he was working."

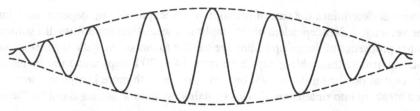

Figure 7.8 Schematic structure of the wave group

tempted to understand why the waves prefer to gather in packets.[98] I am sure that Rayleigh and other physicists of the past gave serious thought to this problem. But a first step to real understanding was made only after 1967, when T. Benjamin and G. Faire carried out extensive theoretical and experimental studies of the stability of periodic waves on deep water. They showed that simple periodic waves are unstable and tend to break up in small groups of waves. During the next year, V.E. Zakharov derived nonlinear equations describing such groups and later found a realistic mechanism for their formation. Very soon, it became clear that the groups of waves all have properties of solitons.

This is a new type of soliton, not yet discussed in this book. At first, it would appear that these solitons are similar to modulated radiowaves or optical pulses. But after short reflection, we see that this is a superficial analogy. Indeed, the electromagnetic waves (at least, in vacuum) are transmitted without dispersive distortions, while wave packets on deep water will diffuse rapidly due to the strong dispersion (recall that $v = \sqrt{g\lambda/2\pi}$). The only obstacle to this diffusion is provided by nonlinearities. So, the groups of waves exist because of a balance between dispersion and nonlinearity, like Russell's solitons.

The story of this new soliton is also intricate. In fact, physicists have met them earlier in nonlinear optics, and very similar equations were known for some time in the areas of superfluidity and superconductivity. We will not go into a detailed discussion of this here.

The soliton, schematically drawn in Figure 7.8 is usually called the *"envelope"* *soliton*; we shall also call it the *"group" soliton*. The word "envelope" is to remind us that the familiar soliton shape appears only in the imaginable enveloping curve; it is shown in Figure 7.8 by hatched lines. The waves under this curve move with a velocity that differs from the velocity of the soliton (a wild life is hidden under the enveloping roof!). In a rough approximation, the waves inside the group are monochromatic. The shape of the enveloping curve is given by the familiar formula

$$y(t, x) = y_0/\cosh((x - vt)/l),$$

---

[98]Groups of waves are easy to understand in the linear theory of waves. The real difficulty is to explain how and under what conditions the wind manages to create packets of waves. This is certainly a nonlinear phenomenon. Likewise, no linear theory can explain the peculiar structure of these packs, which will be discussed shortly.

where $2l$ determines the size of the soliton. Its amplitude, y_0, depends on l, but the velocity, v, is independent of the amplitude, as distinct from Russell's soliton. Other properties of the group soliton are similar to that of the KdV-0-Russell solitons. Normally the stable groups have from 14 to 20 humps under the envelope, the central one being the highest one (the groups with more humps are unstable and break up into smaller ones). This explains the sailors' stories about the "tenth wave."

At this point we finish our passing acquaintance with sea waves and solitons. I have only sketchily drawn the real things existing in the ocean. I hope that it is not too bad a sketch. I wish now to introduce to you, in a similar manner, several other solitons. With this in mind, I first summarize the main properties of the solitons discussed above and give a simple classification.

Three Solitons

Let me invite you to a club of "many-faced solitons." Among its members, we have three most famous solitons: the KdV solitons (Russell's solitons), the FK (Sine-Gordon) solitons, and the envelope (group) solitons. They are the most well known for their exceptional mathematical virtues, allowing for a rigorous mathematical treatment. Not less important is that they have many physical incarnations that may be found almost everywhere. They also have numerous descendants, both in mathematics and physics, that are united in so-called *soliton hierarchies*. In short, if you are acquainted with this "Big Trio," you will be quite at home in the soliton society. Thus, let us first recollect what we know of this Big Trio, emphasizing both common features and individual differences of the three most important solitons.

The Russell–KdV solitons exist in physical systems with weakly nonlinear and weakly dispersive waves. When some wave impulse breaks up into several KdV solitons, they all will move in the same direction. The velocity of the individual soliton is proportional to its height, while its length is inversely proportional to the square root of its height. Collisions of solitons have been described in the preceding chapters.

The group soliton is essentially a smoothly modulated monochromatic wave in weakly nonlinear but strongly dispersive media. The length of the soliton is inversely proportional to its amplitude (not to the square root of it!) and its velocity is independent of the amplitude. These quantitative differences from the KdV soliton are not accidental. Indeed, the shape and the velocity of the KdV soliton are defined by a balance between nonlinear and dispersive effects. Such a balance is rather difficult to imagine in the group soliton. In fact, the group soliton is a very complex and unusual object, which is rather hard to understand in qualitative physical terms.

Under certain conditions, both the KdV and the group solitons may be regarded as particles, obeying the standard laws of Newton's mechanics. For example, this is possible if an isolated soliton moves in a weakly nonuniform medium, in which dissipative effects (friction) are small. On the contrary, the internal soliton structure becomes important in collisions with sharp inhomogeneities of the medium. Note that the soliton is slightly deformed even on a small inhomogeneity (like a small local change in the depth), but it is not destroyed and its velocity is not changed after passing the obstacle. Thus, we may approximately describe its interaction with a small inhomogeneity in terms of a particle, interacting with an attractive or repulsive center.

The influence of friction on the soliton is more serious. The KdV soliton gradually decelerates and becomes smaller and longer. This "degrading" of the soliton follows an exponential law: velocity and amplitude are proportional to the exponential factor $\exp(-t/\tau)$, while length grows proportionally to $\exp(t/2\tau)$. The parameter τ is called the soliton's lifetime; it is proportional to the force of friction. Under the influence of only a small amount of friction, the soliton slowly deforms but preserves its identity. If the friction force is large, the lifetime becomes so small that we can no longer speak of solitons.[99] Equally frail is the nature of the group soliton. The only difference is that friction does not change its velocity, and the length grows proportionately to $\exp(t/\tau)$.

Thus, both the KdV and the group solitons are "mortal." Then, why is the Frenkel–Kontorova soliton so strikingly stable? Under the influence of friction, this soliton only slows down and eventually stops and, at rest, it can live "eternally!" For instance, a dislocation can live as long as its home crystal exists. The reason for the immortality of the FK (SG) soliton is the very special nonlinearity $(\sin\phi)$ in the SG equation. For this nonlinearity, the stability of the soliton has a topological nature, that is, there is no other stable solution into which the soliton at rest can be continuously deformed. For this reason, the FK soliton is called a topological soliton. It is the head of a family of other topological solitons. This third of the Big Trio and its descendants are most fascinating theoretically and have interesting applications, which will be discussed below.

Substituting in the FK equation the term $-2(\phi - \phi^2)$ for $\sin\phi$, one can find a solitary wave solution apparently similar to the FK solitons:

$$\phi = \pm \tanh\left(\frac{x - vt}{l_0\sqrt{1 - v^2/v_0^2}}\right).$$

However, these waves are not true solitons. Being topologically stable, they nevertheless do not preserve their identity in collisions. The FK equation with the term $2(\phi - \phi^3)$, instead of $\sin\phi$, gives rise to a KdV-like solitary wave that is also not a true soliton. This doesn't mean that these objects are uninteresting for physics. On the contrary, all of them are respectable members of the soliton club.

---

[99]In mathematical terms, this means that the KdV equation, with added friction terms, is not integrable. With a very little friction, it is almost integrable.

Newer members continue to join the soliton club, and even experts have difficulties in orienting themselves amidst the throng. Of great help in this are the blood relationships between very different solitons. Even apparently dissimilar members of the Big Trio prove, under a closer examination, to be close mathematical relatives. Their kinship is hidden behind a deep and complex mathematics, but their similar profiles suggest its existence at a glance. Their shapes are defined in terms of the same mathematical function, the hyperbolic cosine (for the FK-soliton, you have to recall that the space derivative of its profile $\phi(t, x)$ is inversely proportional to the hyperbolic cosine). The relation between breathers and group solitons is more obscure but, in fact, the small amplitude breather obeys the same equation as the deep water group soliton.

A very interesting example of the relations we have discussed is the phenomenon of self-induced transparency, which brings us to the subject of optical solitons. The normal absorption of light by a substance can be explained in terms of excitation of its atoms by photons. When a light wave passes through the substance, the electrons may be excited from their ground state, with energy E_0, to a state with a higher energy. In this way, the energy of light is transferred to electrons. Let us assume, for simplicity, that the electrons in the atoms have only two states, with the energies E_0 and E_1. If the frequency of the light wave, ν, is such that the energy of photons in it, $h\nu$, is close to $E_1 - E_2$, the absorption will be particularly strong. This phenomenon is called resonance absorption: the substance becomes opaque for light waves having the frequency $h\nu \approx h\nu_0 \equiv E_1 - E_0$.

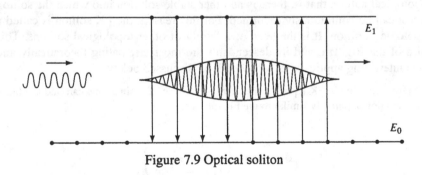

Figure 7.9 Optical soliton

Now, imagine that light passes through this substance in brief pulses. Let us tune the frequency of light to the resonance frequency, $h\nu_0$, and make pulses much longer than the wave period $2\pi/\nu$ (in real experiments this is $\sim 10^{-15}$ seconds and the pulse duration is $\sim 10^{-10}$ seconds). For small amplitudes of the pulses, the crystal will be opaque. However, for large enough amplitude, it suddenly becomes transparent! What happens? Roughly, the following. The front of the pulse throws the electrons to the higher level, with energy E_1, but the shape and the velocity of the pulse are self tuned in such a way that the lost energy is returning to the pulse. This happens at the rear of the reshaped pulse (Figure 7.9).[100]

It is natural to try to use these solitons for transmitting information through optical fibers. By using solitons, one may expect to get a very high speed in transmitting large amounts of information. The recent development of this idea is described in the next section.

Soliton Telegraph

In 'all optical' fiber systems – one without electronic repeaters – a single fiber could transmit as much as 100 Gbit/sec over thousands of kilometers. Such performance would be obtained by using optical gain to overcome fiber loss, and by transmitting the signals as non-spreading, soliton pulses.

Linn F. Mollenauer

The history of the "soliton telegraph" or "telephone" had its start in 1984. A Japanese physicist, Akira Hasegawa, working at the Bell Laboratories in the USA, published detailed results of his computations concerning the problem of the transmission of solitons through optical fibers. He clearly demonstrated that optical solitons, similar to those described above, could travel, say, from New York to Tokyo without significant deformations. In 1988, this theoretical prediction was proved in experiments by L.F. Mollenauer and K. Smith at the Bell Laboratories. They successfully detected soliton transmission over distances in excess of $4000km$. The first experimental demonstration of the soliton transmission in optical fibers goes back to 1980 (L.F. Mollenauer, R.H. Stolen and J.P. Gordon). However, to achieve transmitting solitons over such a long distance, much theoretical and experimental work had to be done and new ideas had to be implemented. Let us look at some highlights of this story.

Attempts to use optical pulses for transmitting information started as soon as good enough optical fibers had become available (about 20 years ago). Due to the small wavelength of the light waves, this way of communication must be very powerful and inexpensive. It is no wonder that it is by now widely used, even in transatlantic communications. One fiber can transmit several hundreds of megabytes per second, and a cable, consisting of many fibers, can transmit several gigabytes per second. This is impressive enough, but the potential of the optical fibers is immensely greater!

It is not difficult to understand why a simple use of optical fibers does not allow us to realize this potential. Even if you use a superb quality fiber, the optical signal in it gradually fades and diffuses due to damping and dispersion. For this reason, one has to install into the optical transmission line some regenerators of

---

[100]It turns out that the shape of the pulse inside the crystal can be derived by solving a FK-type equation, although the pulse is a typical group soliton. This is one of the manifestations of the close relationship between all types of solitons.

the pulses. The regenerator consists of a receiver (detector), transforming optical signals into the standard electronic ones, of an amplifier of the electronic signals, and of a transmitter, which transforms the amplified electronic signals back into optical pulses. The maximal transmission capability of the regenerators is about one gigabit, while the optical fiber can transmit up to one hundred gigabits. In addition, the regenerators should be installed at distances not larger than 50–100 kilometers. So, in long lines, there must be dozens of regenerators, and this increases the probability of errors and breakdowns in communication. In short, the complex structure of regenerators is at variance with the fundamental simplicity of optical fibers. Clearly, one has to look for simpler and more fundamental ways for feeding weak optical signals.

A natural idea is to transmit optical solitons. Indeed, due to their extraordinary stability, it must be much easier to feed them than ordinary optical pulses. To explain the relevant ideas, we first briefly digress into some phenomena of nonlinear optics. Normally, nonlinearity in optical phenomena is very small. For instance, the refractive index in standard optics may be considered as independent of the optical signal amplitude. However, this is not true in laser optics.

If a strong laser pulse is going through the optical fiber, one expects that its refraction index will depend on its amplitude. The reason is that the electric field of the pulse may polarize the molecules of the fiber and thus change the refractive index. The larger the amplitude, the larger are both the field and the polarization. The effect of changing the refractive index in strong electric fields was discovered in 1875 by a Scottish physicist, John Kerr (1824–1907), and it is called the *Kerr effect*. He showed that the increment of the refractive index is proportional to the square of the electric field, but the proportionality coefficient is not large (it is not quite easy to polarize the molecules!). In fact, to clearly detect the effect, one should apply an electric field not less than $10000V/cm$. In normal waves of light, the electric field is not strong enough to significantly polarize the molecules, it is usually less than several hundred V/cm. This explains why the refractive index is normally independent of the amplitude. However, in the laser beam, the electric field may be as large as one hundred million V/cm, which is comparable to the internal electric fields in the molecules. Accordingly, the laser beam may strongly polarize the molecules.

This nonlinearity results in several interesting effects. One of the most remarkable is the *self focusing of the laser beam*. Normally, the cross section of a beam of light in glass grows gradually. However, it may happen that the laser beam, in some types of glass, is focused into a thin string. A qualitative explanation of this effect is very simple. The electric field is maximal on the axis of the beam and so is the refractive index. Recalling the refraction law, one can understand that the outer parts of the beam must be attracted to the axis. The exact theory of this phenomenon is rather complicated. It was first given, in 1971, by Russian physicists A.B. Shabat and V.I. Zakharov. They showed that the waves of light in nonlinear media can be mathematically described by the so called *nonlinear Schrödinger equation*. The same equation describes the envelope solitons.

The immediate consequence of this result is that there must exist optical envelope solitons.

In 1973, A. Hasegawa and F. Tappert applied the nonlinear Schroedinger equation to laser beams in optical fibers and found conditions that allowed for the creation and detection of solitons. Corresponding experiments were performed by Mollenauer, Stolen and Gordon. They transmitted solitons over distances close to 1 km. For larger distances, one has to supply the solitons with energy, to compensate for travel losses.[101] In the best available optical fibers the energy loss is approximately 5% over $1km$. It follows that the energy of the soliton at the distance L is $E_0 \exp(-0.05L)$, where E_0 is its initial energy. So at the distance $L = 20km$, the energy is $\approx E_0/2.72$.

One may suggest different ways of compensating these energy losses. The simplest is to inject laser light into the fiber every 20–40 kilometers. Of course, the frequency and the intensity of the laser must be tuned so that the molecules of the fiber can pick up some part of the energy, stored in the laser beam, and quickly deliver it to the soliton. This resembles the mechanism of the self-induced transparency, but the molecules are excited by the external energy source (laser) instead of the soliton itself. This difference is crucial for transmitting solitonic signals, because now "amplifying" the solitons becomes possible.

The real mechanism of supplying solitons with additional energy is closely related to the *Raman–Mandelshtam* effect, which was discovered in 1928 by Indian physicists Ch. Raman and K. Krishnan and, independently, by Russian physicists L.I. Mandelshtam and G.S. Landsberg (it is often called the Raman effect). The essence of the effect is that the spectrum of the scattered (diffused) light is changed by its interaction with the molecules of the medium. Roughly speaking, the incoming wave is modulated by vibrating molecules. These vibrations are excited by the wave, but the frequency of the vibrations depends only on properties of the molecules.

The mechanism of amplifying solitons is more complicated and we cannot go into details. To understand the concept of the soliton telegraph, it is sufficient to know that the molecules of the fiber serve as mediators in the energy exchange between the soliton and the external laser beam. One can arrange this exchange so that the energy received by the soliton from the beam will exactly compensate for the losses. Then the soliton will travel over very long distances preserving its original shape and velocity. Of course, after many activations of the energy supply, some errors will gradually be accumulated, and they will result in random changes of the velocities of the solitons (the changes in their shape are less important). Thus, for very large distances, the information encoded in the sequence of the solitons may be distorted.

There are also restrictions on the speed of transmitting information. The main one comes from the fact that the soliton has a finite width (length), and there must

---

[101]The velocity of the optical soliton is independent of its energy, but its amplitude diminishes and its width grows (their product is constant).

be enough space between two successive solitons. The minimal duration of the soliton pulse is of the order of 1 ps (1 ps = 1 picosecond = 10^{-12} sec). If the intervals between two successive pulses are 10 ps, transmitting 1 bit of information requires 10 ps and the transmission speed is 10^{11} bit/s = 100 gigabits/s. This limit is difficult to overcome, but it is a very high speed, 100 times faster than the transmission speed in the standard optical fiber communication. In addition, soliton communication is more reliable and less expensive.

In the experiments mentioned above, all these expectations were confirmed. Solitons were transmitted over distances exceeding 4000 km without significant distortion of their shape and velocity. The next step must be a practical realization of the soliton telegraph. This will be the first example of the wide technical application of solitons, on the same scale as telegraph, telephone, radio and TV. Most probably, optical solitons will also be used in computers with optical hardware.

It is worth noting that the story of optical solitons is a very instructive and remarkable example of the relations between fundamental science and engineering. After the fundamental work of Zakharov and Shabat (1971), the idea of solitons in optical fibers appeared shortly thereafter (1973). Just at that time the technology of making high quality fibers was developed while lasers were the standard equipment of physics laboratories. At the same time, the idea of using the Raman–Mandelshtam effect for amplifying signals in optical fibers was formulated. In 1980, solitons in fibers were detected, and three years later it became clear that the Raman–Mandelshtam effect might also be used for feeding solitons. After five years of computations and experiments, the technical realization of the soliton telegraph idea was achieved. Now it is a problem for engineers, technicians and businessmen.

The Nerve Pulse —
Elementary Particle of Thought

> Is not Animal Motion performed by the Vibrations of this Medium,
> excited in the Brain by the power of Will, and propagated
> from thence through the folid, pellucid and uniform Capillamenta
> of the Nerves into the Muscles, for contracting and dilating them?
>
> Newton

As you know from Chapter 3, the basic concepts of electric pulse formation in nerves and its transmission through nerve fibers were formulated early in this century. They, however, were not confirmed in experiments on living nerves. There were several obstacles to these experiments but, probably, the most important one was that the nerve fibers (axons) are normally very thin. For instance, the diameter of nerves in mammals is less than 20 micrometers; the most popular frog's nerves are not much thicker, only 50 micrometers.

So it happened that a real study of nerve fibers and experiments with transmitting electric pulses through them began in 1936, after J.Z. Young discovered the squid giant axon. Some nerves of these strange creatures are unusually thick, up to 1 millimeter in diameter, and this unique property of their nerve system was of great value for science.

It is not difficult to understand why squids are so exceptional. The transmission speed of nerve pulses grows with the diameter d of the central (inner) part of the axon. However, it grows slowly, approximately proportionally to $d^{1/4}$. To survive in an unfriendly environment, one has to react to signals of danger as fast as possible. In particular, the speed of the nerve pulses in the axons must be high. The simplest way to achieve a high speed of pulses is to make your nerves very thick. This is the method used by squids in their evolution, and this is why these strange creatures have survived up to our time. However, making the nerves very thick is not the only possibility. In fact, it is not the best solution of the problem. Life is more ingenious than that, and many more effective constructions of the nerves were "invented" by the evolution process. In particular, mammals have a different construction of axons, which supports somewhat different mechanisms of transmitting nerve pulses. Most of the mammal's (and human's) nerve fibers have an isolating cover (membrane) surrounding the inside (core) fiber. This construction resembles a standard TV cable and makes it possible to achieve high speed in fairly thin nerves. The result is that the speed of the pulse in the thick nerve of the squid is 25 m/sec, while in our nerves, which are 50 times thinner, it is up to 100 m/sec.

Now, let us return to the modern story of the nerve pulse. Young's discovery enormously accelerated nerve pulse study, and before mid-20th century the main properties of pulse propagation were experimentally established. This prepared the way for a real theoretical breakthrough — in 1952 two English physiologists, A. Hodgkin and A. Huxley, published a series of brilliant papers in which they proposed the first reliable theory of the nerve pulse. This theory soon received worldwide recognition and, in 1963, Hodgkin and Huxley won the Nobel Prize in Medicine. We will not go into a detailed description of their theory or of nerve fiber structure and function. A brief acquaintance with the main facts and ideas will suffice for understanding the soliton-like nature of nerve pulses.

Many experiments have shown that the shape and velocity of the pulse are quite independent of the nature and strength of the original nerve irritation. If the irritation is very strong, the nerve responds by a burst of nearly identical pulses. There will be no response at all to a very weak irritation; the minimal strength of the irritation is called the "threshold" irritation strength. All this resembles the decay of a big hump on the water surface into smaller KdV solitons. The only significant difference is that the nerve pulses are identical and travel with the same speed (recall that the KdV solitons can move with arbitrary velocity and their shape depends on the velocity). The simplicity and expediency of this design for transmitting information in living creatures are very remarkable. Each pulse carries just one unit of information ("yes/no" message) and so the "receivers"

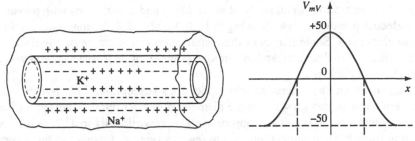

Figure 7.10 Nerve pulse

need only count the number of these "elementary particles" of information in a given time interval.

Now, let us try to understand the mechanism that transforms irregular and diverse signals, coming from the outside world, into regular sequences of identical solitary waves. We have first to stress that the nerve pulse is not the electric current pulse, because the nerve fiber is a very bad electric conductor. Indeed, a thin 1 meter long nerve fiber has the electrical resistance of a copper wire having length of the order 10^{10} km. Fortunately, for transmitting voltage pulses, there is no need to have good conductance; we will see in a moment that nerve pulses are in fact better described as voltage pulses. This was first clearly understood by an American physiologist, K.S. Cole, in 1949.

The simplest nerve fiber consists of a core, which is surrounded by a membrane (see Figure 7.10). There is an ionized liquid inside and outside the membrane (internal and external plasmas). The outside plasma has an excessive amount of sodium (Na^+) and chlorine (Cl^-) ions, which are products of the dissociating salt ($NaCl$). The inside plasma has an excess of potassium ions (Ka^+) as well as of negatively charged ions of some organic molecules. The membrane is permeable only to the inorganic ions Na^+, Ka^+ and Cl^-, but the large organic ions cannot penetrate it.

Now, in a quiet state of the nerve, there is a dynamical balance among the ions penetrating through the membrane in both directions. In the corresponding equilibrium state, the core of the fiber has a net negative charge, while the outer plasma is positively charged; the electric potential between them is approximately $50 mV$. When the nerve is excited by a strong enough external perturbation, the membrane reacts by allowing the sodium ions to go inside. The result is that near the location of external irritation the potential changes its sign. This influences the adjacent parts of the membrane, and so the voltage pulse propagates along the fiber (see Figure 7.10). We see that the nerve pulse is an electro-chemical reaction traveling along the fiber.

Such a mechanism may work only if there exists a nonlinearity that amplifies large deviations from the quiet state and suppresses the small ones; otherwise the pulse could not exist. Indeed, let us imagine, for a moment, that everything is linear. Then the front of the pulse will flatten because the diffusion of the ions

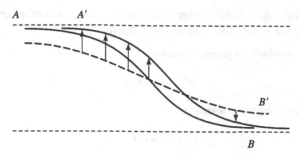

Figure 7.11

through the membrane is proportional to the potential. This is quite similar to the diffusion of a drop of ink in water. In fact, both processes are mathematically described by the same linear differential equation, which is called the *diffusion equation*.

The reason why the nerve pulse is not diffused is that the dependence of the membrane permeability on the potential is not linear, and the nonlinearity acts against diffusion. This is schematically depicted in Figure 7.11. If there were no nonlinearity, the deformation of the pulse would result in changing its shape to the dashed line. The action of the nonlinearity on the shape is symbolically represented by the arrows. The balance between the effects of diffusion and nonlinearity results in moving the pulse without changing its shape.

Propagation of the nerve pulse is mathematically described by the *nonlinear diffusion equation*. The balance between nonlinearity and diffusion incorporated in this equation is analogous to the balance between nonlinearity and dispersion in the KdV soliton. However, there are significant differences. The main difference is that the shape and velocity of the nerve pulse do not depend on the initial conditions. In fact, in a given fiber, the velocity of the pulse and its shape are fixed. In addition, the distance between successive pulses in a burst of pulses, generated by a strong irritation of the nerve, is also fixed. Clearly, it is a most effective and conceptually simple mechanism for transmitting information. The problem of supplying the energy for compensating inevitable energy losses is also solved in a simple way. Any additional energy coming from the electrochemical reactions just keeps all these parameters fixed. The reason is that the solutions of the nonlinear diffusion equation, corresponding to different signals, are stable and well separated in energy. For this reason, there is a balance between the energy losses and supplies (similar to the balance in a clock, where the energy supply does not change the period or the amplitude of the pendulum oscillations).

The nonlinear diffusion equations (with different kinds of nonlinearities) are interesting in many other branches of science. The first serious investigation of a typical nonlinear diffusion equation, describing solitary waves, was undertaken by A.N. Kolmogorov, I.G. Petrovskii and N.S. Piskunov, in 1937. They published a beautiful paper (also related to a biological problem), in which they showed that nonlinearity and diffusion effects may compensate each other and produce a stable solitary wave with fixed shape and velocity.

In fact, they considered a generic equation,

$$u'' - \dot{u} = F(u) \, ,$$

and found its travelling wave solutions,

$$u(x, t) = u_S(x - vt).$$

They applied phase portrait techniques for determining u_S, and derived explicit expressions for it with some simple functions $F(u)$.

Actually, their equation also was a good mathematical model for nerve pulse propagation, but nobody had noticed that. Reviewing the history of the nerve pulse, A.C. Scott remarked: "This uniquely important contribution was completely overlooked by electro-physiologists in the U.S.A.; indeed it is not even noted in the otherwise exhaustive bibliography of the book by Cole (1968). . . . the difficulty may have been the assumption by most mathematicians that the diffusive and non-propagating behavior of linear diffusion equation would carry over to the nonlinear case." Physicists proved to be more receptive to this innovation. A year later, the Russian physicists Ya.B. Zeldovich and D.A. Frank-Kamenetskii successfully applied the results of Kolmogorov, Petrovskii and Piskunov to the theory of burning. Starting from this development, Zeldovich proposed (1939) a theory of the *shock waves of burning*, known as *detonation waves*.

The solitary wave of flame (the burning or combustion wave) normally moves slowly. For instance, if you light a fire at one end of a tube filled with a combustible gas, a solitary wave will start propagating along the tube. Its velocity is usually small, about ten times less than the speed of sound in the gas. However, if you light the fire with a very strong spark (please don't try to do this, it is really a very dangerous experiment!), *detonation* may result. The detonation is, in fact, a shock wave of flame (burning). The velocity of this wave is very large, several times larger than the speed of sound. The shock waves of flame are also called explosive waves or detonation waves. The foundation of the theory of these waves was laid in the year 1939 by Zeldovich. He showed that the speed of the detonation wave is defined by the energy, released in the chemical reaction of combustion, and he showed how to calculate this speed for particular waves. The most important thing is that the detonation wave, like the normal wave of flame, has a definite velocity, determined by the reaction. Recall that the usual shock waves in air may have different velocities and gradually die down.

READER: I understand that the waves of flame and detonation waves somewhat resemble the nerve pulse. What about possible existence of a shock nerve pulse?

AUTHOR: I never heard of anything like this. Not being an expert in this field, I cannot give a more definite answer to your interesting question, but I can refer to my own experience. At least twice in my life, I experienced a very strong acceleration of my reactions to external dangers. It appears that, at the moment of a deathly danger, time may be "stretched" and you can think and act with a speed much higher than normal. This is a very unusual and strong experience: "There is an ecstasy in battles and at the edge of an abyss" (A. Pushkin). Perhaps we have some other, much faster mechanism for transmitting information through our nervous system, which, by the way, requires much more energy to function and so, normally, is not working (after a short period of ecstasy you feel completely exhausted). I have no idea of how this acceleration may work.

Let us stop at this point and leave the subject to professional biologists, chemists, physicists, mathematicians, and engineers. Why engineers? Because the nervous system is a very interesting communication system that can serve as a pattern for telecommunication and computer engineers. Of course, before trying to model the nervous system, much work has to be done. Uncovering the nature of the nerve pulse is only a first step on this path and, in addition, some details are not quite clear even with the single pulse. A much more complex problem is functioning of the nervous system as a whole.

How do nerve pulses determine the movements of our muscles? How do pulses generate other pulses and finally produce feelings and images of the outer world in our brain? Are these "elementary particles of thought" directly related to our real thoughts and feelings and, if they are, what is the precise relation? There are many other simple questions to which modern science has no definite answers. But the important thing is that these questions are gradually making sense, and we are starting to look for scientific keys to these questions. We do not know what will be the final answers but, even now, it is quite clear that solitary waves are important instruments of life on a very deep, fundamental level.

Vortices – Everywhere

> But they live, they live in N dimensions
> Vortices of wills, cyclones of thoughts, all those for
> Which we are ridiculous with our childish vision,
> With ability to only step lengthwise.
>
> Valerii Bryussov

Up to now, we have really taken only lengthwise steps. All solitons and solitary waves described above were essentially one-dimensional objects (i.e., with $N = 1$). This means that they depend on only one coordinate.

For instance, a wave coming to the shore may be approximately regarded as one-dimensional. Circular waves generated by a stone thrown into water exemplify two-dimensional waves, while spherical light waves from a lamp are three-dimensional. Modern physicists are also interested in waves travelling in higher-dimensional spaces. It is no wonder that physicists and mathematicians are trying to discover solitons and soliton-like objects in N dimensions.

Few N-dimensional solitons are known today. The most important and systematically studied are two-dimensional solitons similar to the KdV solitons. They are generated by two-dimensional nonlinear waves, the equation for which was first suggested in 1970 by B.B. Kadomtsev and V.I. Petviashvili.[102] Several years later, two types of the soliton solution of their equation were discovered. One describes collisions of the usual one-dimensional KdV-like solitons, moving on the surface in non-parallel directions (a simpler but similar problem is the reflection of a soliton from a wall, as first observed by Russell). Another one is the real two-dimensional soliton, having the shape of a lump, the height of which is decreasing in all directions from its center. Figuratively speaking, one may say that the one-dimensional soliton is like a mountain ridge, while the two-dimensional soliton resembles a single hill on a valley.

Sometime later, two-dimensional generalizations of group solitons on water surfaces were discovered, and today several interesting solitons in three and more dimensions are known. We will mention some of them below.

Let us now discuss interesting solitons related to vortices. What is the meaning of the word "related"? This relation is twofold. On the one hand, there exist solitons "living" on vortices; vortices provide auspicious conditions for solitons. On the other hand, vortices, as well as more complex formations of vortices, may also be regarded as many-dimensional soliton-like objects. We first discuss the simplest solitons living on vortices.

Around the crater of the vortex on the water surface in a bathtub, one can sometimes observe spiral waves, which look similar to the spiral sleeves (branches) of the galaxy discovered by Lord Ross (recall Figure 3.8). In fact, both phenomena have the same nature, although they are very much different in space and time scales. The spiral sleeves are formed by waves in interstellar gas. All the details of the complex processes that lead to forming these sleeves are not yet completely clear, but a natural and popular assumption is that the visible thickenings of the interstellar gas are created by gigantic solitary waves on the background, formed by the galactic vortex. This is supposed to be similar to spiral waves around the vortex in the bath.

You may meet another interesting soliton while bathing. One Russian poet, Leonid Martynov, described it in his verses (I am quite sure that he had no knowledge of solitons):

> Into the outlet of my bathtub
> Water is gliding. Spiralling
> Exactly like bindweed — tornadoes
> That travel all over the world.
>
> Leonid Martynov (1905–1980)

Please take notice of the words "gliding" and "spiralling" (my translation into English is literal; the Russian version is more colorful). These words give a rather precise qualitative description of waves and of solitons on the vortex "stem" (or

---

[102]They considered the evolution of weakly nonlinear waves in dispersive media.

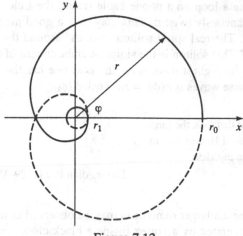

Figure 7.12

"leg"). As distinct from waves on a rubber band, these waves glide like a snake, or a bindweed. A natural question is to ask why they are "spiralling."

The answer is very simple. Let us recall that vortex rings (see Chapter 4) move perpendicular to their plane and that the velocity of this movement is inversely proportional to the diameter of the ring. With this knowledge, it is not difficult to understand that the bending part of the vortex line (the stem of the vortex) will move perpendicular to the plane in which it lies. As the velocity of this movement is proportional to the curvature[103] of the vortex line, the bending curve is spiralling. One may say that the waves of bending are necessarily spiralling. The mathematical consideration of such waves shows that there must exist solitons that might be called "spiral solitons."

If you look at the spiral soliton from the top, along the original direction of the vortex line, you will see the picture shown in Figure 7.12. The picture shows the projection of the vortex line on the $x - y$ plane; the dashed curve represents the part of the line that lies below this plane. The points on the curve in the $x - y$ plane satisfy a very simple equation: $r = r_0/\cosh(\phi/\phi_0)$; z is approximately proportional to ϕ.

Figure 7.12 is a snapshot of the projection. In reality, the spiral soliton is always in uniform motion along the z direction. Correspondingly, its projection on the $x - y$ plane uniformly rotates. The velocities of these uniform motions are inversely proportional to the amplitude of the soliton, r_0 (this corresponds to the inverse proportionality of the velocity of the translational movement of the vortex ring to its diameter). You will get a better idea of the shape of the spiral

---

[103]The curvature of the curve at a point is defined as follows. Let us imagine a circle most closely fitting the curve at the point (it is called the tangent circle). The inverse radius of this circle is called the curvature.

soliton if you make a loop on a phone cable (recall the Euler soliton) and then stretch and simultaneously twist it. This may give a good model for the central part of the soliton. The real spiral soliton "twines" around the vortex line, quite like the bindweed! This soliton is the simplest embodiment of the group soliton. The amplitude of the highest wave is r_0, the next one has the amplitude r_1, etc. The envelope of these waves is $r(\phi) = r_0/\cosh\phi/\phi_0$.

> And the cloud that took the form
> (When the rest of Heaven was blue)
> Of a demon in my view.
>
> Edgar Allan Poe (1809–1849)

Spiral solitons of much larger dimension may be observed in tornadoes. The tornado is usually generated by a vortex inside a black cloud. For some unknown reasons, the end of a big enough vortex sometimes sets down like an elephant's trunk. When this trunk touches the surface of water or soil, a water spout or dust storm arises, having the shape of a giant relatively slowly moving pillar (its velocity is 30–70 km/h). This is what is usually called a tornado (from the Spanish word *tronada*). The height of the pillar may be up to 1–1.5 km, and the diameter might be up to 1 km (tornadoes in the sea have smaller diameters, several dozen meters). The inner parts of the pillar are rotating very fast; the velocity of the whirling wind inside might be as high as 300 km/h! In addition, the atmospheric pressure inside is very low. The combination of rarefaction of the air with the high speed of rotation defines the destructive potential of the tornado. The violent whirl captures small things on the surface and carries them into the cloud. Later these things may fall back with rains (for instance, so-called fish rains are well known to sailors). The sudden jump in pressure when a tornado hits a building may result in the destruction of the building due to the large difference between outside and inside pressure.

Let us return to the spiral solitons in tornadoes. Looking at a tornado (I saw many of them on the Black Sea), one can often observe the smooth oscillations of its trunk. These oscillations may give origin to spiral solitons similar to those on the whirls in the bathtub. The most interesting question is why and how these trunks leave the cloud and finally form the tornado. I think that the spiral oscillations and solitons play a significant role in these processes but, to the best of my knowledge, there is no quantitative theory describing them (on the contrary, the structure of a well-developed tornado is fairly easy to describe). This makes prediction of tornadoes rather a difficult problem.

There are other solitons living on vortices. Quite harmless and simple solitons may exist on the boundaries of vortex regions. To describe a vortex region, imagine a flow of water in a plane and suppose that, in some finite region, the flow is whirling around, see Figure 7.13. This means that, in any point of this region, the velocity of water is ωr (in the rest system of its center, O), where r is the distance from the point to the center. In other words, if water in this region had suddenly

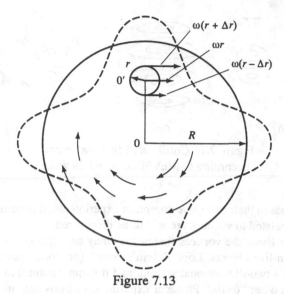

Figure 7.13

froze, the corresponding rigid circle would rotate with constant angular velocity
$\omega = 2\pi/T$, where T is the period of the rotation.

Although it might look somewhat puzzling at first, it is easy to show that each
drop of water inside the vortex region is rotating with the same angular frequency.
To see this, consider the motion of a small circular region with center O' and
radius Δr (see Figure 7.13). The velocity of the point O' (relative to the center
O) is ωr, where $r = OO'$. Now, it is not difficult to find that in the rest system
of the center of the drop, O', the drop is rotating with angular frequency ω (e.g.,
the nearest to O point of the drop has the velocity $-\omega\Delta r$ relative to O', while the
speed of the farthest one is $+\omega\Delta r$). In this sense, all the points inside the vortex
region are equivalent; it is said that water inside the vortex region is in a state of
uniform whirling motion.

Now, on the border between the vortex region and the rest of the water, where
there is no whirling (vorticity), there may exist waves and solitons that change the
shape of the vortex region. Such a change is shown in Figure 7.13 by the dashed
curve. As the soliton moves along the border, the vortex region will look to be rotating. To observe these solitons in natural conditions[104] or even in experiments is
not easy (I never have seen them), but they were found in numerical experiments.
In 1978 Gary S. Deem and Norman J. Zabusky discovered rotating vortex regions

---

[104]Large scale vortex regions are known as tropical hurricanes or typhoons. The diameter of these
giant vortices can be as large as 100-500 km. Although they move slowly along the earth or sea surface
with speed less than 15-20 km/h, the winds on their periphery may be terrible, having speeds up to
$250 km/h$. In the center of the typhoon the winds are weak; there even may be a dead calm. However,
this is delusive. It is no pleasure to be at the center of the typhoon, because of a very low atmospheric
pressure (it can be more than 10% below normal).

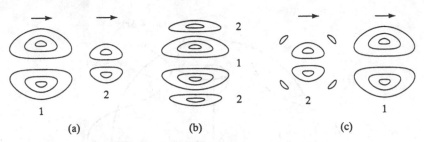

Figure 7.14 Collision of two vortex pairs
according to McWilliams and Zabusky

of diverse shapes in their extensive computer experiments. It is quite possible that other solitons related to vortices are still to be discovered.

On the other hand, the vortices themselves may be regarded as solitons or, at least, as soliton-like objects. Lord Kelvin's "oval" (or, vortex pair, recall Figure 3.4) looks like a two-dimensional soliton, and it is quite natural to investigate its collisions with other "ovals." Physical experiments of this sort are very difficult and the word "investigate" here implies computer experiments (analytical solution is hardly possible). The result of one of these computer experiments is schematically shown in Figure 7.14. The bigger and faster pair runs down the smaller pair (see *a*). The next picture (see *b*) shows that the bigger pair splits the smaller one into two separate vortices. The last picture (see *c*) shows that, after the collision, both pairs are almost unchanged (except small new eddies around the smaller pair). This and other, more complex computer experiments clearly demonstrate that Kelvin's pairs behave like real solitons, at least approximately.

The most familiar vortices, described above, live in liquids and gases. We also mentioned vortex-like structures in galaxies. There exist vortex-like objects in the atmospheres of the planets. A famous example is the Great Red Spot on Jupiter. Some scientists believe that this is a soliton, having a vortex-like structure. Very complex vortices are known to exist in the atmosphere of the sun. The well known spots on the sun are believed to be the ends of vortices, like those shown in Figure 3.5. Most probably, many closed vortex rings also exist inside the sun. They have a much more complex structure than do hydrodynamic vortices as they exist in a charged liquid (plasma) and thus involve strong magnetic fields and electric currents. Even more complex vortices are described in the next chapter. The reader will find interesting material there from the life of vortices in superfluids, superconductors and in high energy physics. Some of the vortices are well established and were experimentally studied in physics laboratories, while others are only theoretically predicted. Not all the predictions will prove to be true, but I will include even speculative ideas to show the reader a picture of scientific work in its real perspective.

Chapter 8
Modern Solitons

Although many of the solitons described in previous chapters were discovered and studied in the 20th century, they could have been discovered and understood by physicists of the previous century. What I have in mind, of course, is the distinction between classical (19th century) and quantum physics (20th century). It happened that the quantum revolution in physics started at the frontier of the two centuries. At the end of 1900 the German physicist Max Planck (1858–1947) boldly introduced the notion of quanta of energy in the processes of emission of light by atoms. He also introduced a new fundamental constant in physics, the Planck constant h. The significance of this revolutionary step was not recognized until much later. Planck himself did not realize that he had started one of the greatest revolutions in physics.

Five years later, A. Einstein introduced the concept of light quanta endowed with the energy $E = h\nu$, where h is the Planck constant and ν is the light wave frequency. It took almost ten years to realize that the light quanta are also endowed with the momentum $p = h/\lambda$. The particle interpretation of the light quantum was finally accepted by the physics community only after the experimental proof of the laws of energy and momentum conservation in collisions with electrons; this was done in 1922 by Arthur Compton (1892–1962)). The term *photon* was introduced in 1926 by G. N. Lewis. Photons are very special particles because they have zero mass. The wave particle duality for massive particles was introduced by Marquis Luis de Broglie (1892–1987) in 1923. He suggested that to any particle there corresponds a *quantum wave*, the length of which is $\lambda = h/p$. At first,

A.T. Filippov, *The Versatile Soliton*, Modern Birkhäuser Classics,
DOI 10.1007/978-0-8176-4974-6_8, © Springer Science+Business Media, LLC 2010

this looks like a trivial rewriting of the above formula for the momentum of the photon but its meaning is quite different and says that quantum waves are not usual waves. A complete understanding of this came later, when the quantum theory was formulated. The wave nature of electron beams was experimentally proved in 1927 by Clinton Davisson (1881–1958) and Lester Germer (1896–1971) and, independently, by George Paget Thomson (1892–1975), who observed a typical diffraction picture for the electrons scattered by a crystal lattice (similar to the diffraction picture produced by the light waves).

Meanwhile the quantum revolution developed on the side of the theory of the atomic structure revealed in their spectra. After the milestone papers of Planck and Einstein, the greatest breakthrough came from Niels Bohr (1885–1962). In 1913 Bohr developed the first quantum model of the hydrogen atom. This triggered the search for systematic quantum laws that would replace the laws of classical mechanics, and resulted in the successful formulation of *quantum mechanics* in 1925–1926 by the German physicist Werner Heisenberg (1901–1976). At the same time, Austrian physicist Erwin Schrödinger (1887–1961) formulated his wave equation for the electron waves in the atom, which gives an alternative explanation of quantum laws. The physical interpretation of the new theory was clarified by the German physicist Max Born (1882–1970). Born proved that quantum waves are really very different from classical waves, with which we dealt in previous chapters. In general, these waves are described by complex functions $\psi(t, x)$, and only the moduli of these functions are measurable physical quantities giving the probability of finding the particle at place x at time t. In other words, these waves are "probability waves."

The mathematical formalism of the new theory was perfected by the English theorist Paul Adrien Maurice Dirac (1902–1984) who proved that the Heisenberg and Schrödinger formulations of the quantum laws are not only completely equivalent, but are in fact two mathematically complementary representations of the same theory. A more general idea of complementarity in the quantum picture of the world was formulated by Niels Bohr at the end of 1927.[105] This was the end of the quantum revolution[106] and the start of the new, quantum era in physics.

Today quantum concepts and methods are used in many branches of the natural sciences, and of course they cover all of physics. It is easy to understand that they are relevant to astrophysics (e.g., quantum cosmology is trying to quantize the whole universe from the beginning of creation) or to chemistry (quantum chem-

---

[105] As a special case, the principle of complementarity involved the complementarity between the wave and particle pictures. Bohr's predecessor was Einstein who in 1909 predicted: "It is my opinion that the next phase of theoretical physics will bring us a theory of light that can be interpreted as a kind of fusion of the wave and emission theory." Here "emission" means the old Newtonian particle theory of light.

[106] All the above-mentioned physicists became Nobel Prize Winners for their contribution to the development of quantum theory: Einstein in 1921, Bohr in 1922, Compton in 1927, De Broglie in 1929, Heisenberg in 1932, Schrödinger and Dirac in 1933, Davisson, Germer and Thomson in 1937, Born in 1954.

istry explains chemical laws by quantum properties of molecules and atoms). It may not be so obvious, but it is true that any deep understanding of living creatures requires quantum mechanics. Thus, molecular biology and molecular genetics widely use quantum results and methods. One may say that quantum theory is relevant to the theory of life processes on a deeper level. Niels Bohr and Erwin Schrödinger foresaw the deep applications of quantum mechanics years ago, and their predictions are gradually being realized.

We all know that the impact of quantum mechanics on mathematics has been enormous. New concepts and even new branches of mathematics emerged through the combined efforts of mathematicians and physicists, and modern theory, trying to find the ultimate laws of Nature, requires the most refined and abstract mathematical methods. No wonder that modern theorists often can't say whether they are solving a mathematical or a physical problem. A large portion of their work is devoted to inventing new mathematical tools or even concepts, and modern theoretical physics is becoming increasingly inseparable from modern mathematics.

To some extent, this is also true for soliton theory, although soliton theory is more directly related to experiments. Moreover, soliton-like solutions are often encountered in modern theories of fundamental interactions. The reason for this is that the deepest and most difficult problems of fundamental interactions are essentially nonlinear. Any detailed description of such things requires using complex and very abstract mathematical concepts and methods. So, I'll only give an intuitive and imprecise picture of some modern solitons, mostly using analogies with simple classical ones.

We shall start our short tour of modern ideas with vortices in quantum fluids. As distinct from the classical vortices, they are, in a certain sense, quantized. For an intuitive understanding, you do not have to know quantum mechanics, but Planck's constant h will enter the main formulae for these solitons. This is a new feature that we never met before.[107] All the solitons we have met above contain so many quanta that the quantum effects are completely irrelevant; we say that they are *macroscopic classical* objects. The solitons we consider below also consist of many quanta and are in this sense macroscopic. But the quantum effects are still very important, and so they are sometimes called *macroscopic quantum* objects (as distinct from the microscopic quantum objects like photons, atoms, etc.). They were first discovered in superfluids and superconductors.

---

[107]In fact, when we discussed self induced transparency, the Planck constant was mentioned when we tried to explain the resonance absorption. However, we might completely forget about it because both the photon energy $h\nu$ and $E_1 - E_2 = h\nu_r$ are proportional to h, and thus the condition of the resonance may be expressed in terms of the frequencies ν and ν_r. The resonance absorption was known in the 19th century, long before quantum physics, and the optical solitons are not quantized in any sense.

Vortices in Superfluids

Superfluidity and superconductivity were not predicted by theorists and puzzled them for decades. We begin with superfluidity, although this was discovered a quarter of a century after superconductivity.

In 1908 the Dutch physicist Heike Kammerlingh Onnes (1853–1926) used a technique for obtaining very low temperatures to liquify helium.[108] Under normal atmospheric pressure it becomes liquid below the *critical temperature* $T_c = 4.2$ K and remains in the liquid *phase* at arbitrary low temperatures (it may be solidified by applying the pressure higher than 22 atmospheres). Thus, at low temperatures helium may exist in three different phases: solid, normal liquid called helium I, and anomalous liquid called helium II. We will now concentrate on the helium II, which is a very strange liquid.

After Onnes died, his cryogenic laboratory in Leiden continued to pursue studies of low temperature physics under the leadership of another Dutch physicist, W.H. Keesom (1876–1956). Keesom concentrated on studying the main physical properties of helium and finally discovered a puzzling effect: namely, below T_c the heat conductivity of helium becomes enormous, practically infinite! This means that heat is transmitted through the liquid helium practically instantaneously.

An excellent experimentalist, the Russian physicist Pyotr Leonidovich Kapitsa (1894–1984) was very much puzzled by this fact. Using his original methods for obtaining liquid helium, he started a systematic search for other strange effects. Thus, in 1938 he had found that the viscosity of the liquid helium suddenly drops at the temperature T_c and below, and becomes at least one million times smaller than above at this temperature. He boldly conjectured that the viscosity was not only fantastically small but exactly zero, and named this phenomenon *superfluidity*. In the same year this phenomenon was independently discovered and studied by the English physicists John F. Allen and A.D. Misener.

Kapitsa was working from 1921 to 1934 in England, first at the Cavendish Laboratory then as the Head of the Mond Laboratory of the Royal Society. There he organized a research in low temperature physics and invented new devices for making liquid helium. In 1934 he came to Moscow for a short visit but was not permitted to return to England. In exchange, the Soviet rulers provided him with means to establish his own institute in Moscow. So he became the head of the now famous Institute of Physical Problems. Rutherford helped him to arrange transfer of the main physics equipment from England to Moscow, and so the interruption in Kapitsa's research was not as long as it could have been. In 1937 Kapitsa invited Lev Landau to become the main theorist of the institute.

---

[108]Like Ernest Rutherford, Kammerlingh Onnes was a fine experimentalist during the first quarter of the 20th century. He and Rutherford paved the way for the new physics; Rutherford — in experiments on the structure of the atom and in laying the foundations for nuclear physics; Kammerlingh Onnes — in discovering quantum properties of matter at low temperatures. Thus, by looking for unexpected phenomena at lowest possible temperatures, he discovered in 1911 the phenomenon of superconductivity — disappearance of electrical resistance in some metals.

Kapitsa and Allen[109] discovered several new and unexpected properties of superfluid helium. Kapitsa organized a systematic search for new effects and strongly stimulated theoretical study of the problem. In 1941 Lev Landau completed a detailed theory describing all observed phenomena and predicting new effects. Some of his ideas and results were anticipated in the theory put forward somewhat earlier by the Hungarian theorist Laszlo Tisza,[110] but his paper became known in Russia several years later. (If you look into the history of U.S.S.R., you could imagine what the years 1937–1938 and 1941 were like for those who happened to live there. In 1938 Landau was arrested by the NKVD and spent a year in jail waiting for the worst. Kapitsa rescued him by appealing to Stalin and Molotov.) Anyway, Landau's theory was more rigorous and detailed (when Landau became familiar with Tisza's papers, he strongly criticized some important points of them), and its influence on the further development of superfluidity and superconductivity was more significant. In particular, he insisted on the quantum nature of superfluidity and immediately realized that there is a deep connection between superfluidity and superconductivity.

According to Landau and Tisza, below the critical temperature T_c, liquid helium behaves as if it consisted of two components — one is a normal viscous liquid while the other has zero viscosity. When the temperature approaches absolute zero all liquid becomes superfluid. The motions of the superfluid component are similar to those of the ideal Euler liquid but there are important differences because quantum effects cannot be neglected in the superfluid component; it is a *quantum liquid*. The hydrodynamics of these two fluids rather successfully explained the main observed phenomena and was subjected to critical experimental tests by Kapitsa and his collaborators. However, Landau was not satisfied and continued to search for a microscopic quantum explanation.

He especially pursued ideas about *rotons* — quanta of rotary motion in the superfluid component. From the quantum (microscopic) point of view all motions in superfluid liquids must be transmitted by *quasiparticles*, i.e., quanta of excitations behaving like particles. Typical quasiparticles are *phonons* – quanta of sound excitations having the same dispersion law as photons. Landau proved that superfluidity is impossible if only phonons exist. For this reason he introduced new hypothetical quasiparticles, quanta of rotational motions, which he called *rotons*. Unfortunately, it was not clear how to prove the existence of these particles from first principles, i.e., in the framework of a microscopic model of superfluidity.

---

[109] Allen came to the Mond Laboratory in 1935. After WWII he was Professor of the Saint Andrew University.

[110] From 1935 to 1937 Tisza was working at the Kharkov Physics Technical Institute where Landau headed the theory group. At that time Landau was very actively working on superconductivity and his influence impacted on other theorists. By the way, I would like to remind you that I am not attempting here to give a full and objective history of physics and much less to be a judge in matters of priority. I am only trying to tell you what I know and guess. The history of superfluidity and superconductivity is a very complex thing and, to my knowledge, it is not yet completely established.

The first microscopic explanation of superfluidity was advanced in 1947 by N.N. Bogoliubov. Instead of fluid he studied a weakly non-ideal gas, and constructed a rigorous quantum theory in this simplified case.[111] Thus he explicitly demonstrated that in this gas there exist excitations with the dispersion law satisfying the Landau criterion for superfluidity. This was a breakthrough in the problem of superfluidity and Bogoliubov's approach to it strongly influenced further development of superfluidity, superconductivity and of quantum statistical physics in general. He himself also tried to explain superconductivity by applying his methods to the "electron liquid" (to obtain superfluidity of the electron liquid in superconducting materials). This attempt was not successful because the electron liquid is fundamentally different from superfluid helium.[112]

A breakthrough in the theory of superconductivity occurred almost ten years later, in 1956, when a young American theorist Leon Cooper proposed a rather paradoxical idea about the structure of the electron liquid in superconductors. His proposal was the following. Although Coulomb forces between two electrons repel one another, there could exist an attractive force between them due to their coupling to atoms in the crystal lattice (phonon exchange forces). These forces may happen to be strong enough for making a bound state of two electrons. This bound state is a quasiparticle, called the Cooper pair (note that electrons as well as phonons are quasiparticles and that the Cooper pair is a very loosely bound system of two electrons, not well localized in space). Now, under certain conditions at a very low temperature, the weakly non-ideal "gas" of the Cooper pairs may become superfluid and this explains the phenomenon of superconductivity. These ideas were given a mathematical embodiment in the paper soon published by Cooper with John Bardeen and John Schrieffer, experienced theorists and experts in superconductivity. Starting from Cooper's idea Bogoliubov immediately adapted a method used in superfluidity and gave an independent derivation of superconductivity. Although the original model of Bardeen, Cooper and Schrieffer was criticized for some unrealistic features, subsequent work soon proved that the main result — the existence of superconductivity — was correct.

We cannot go into a detailed explanation of superfluidity and superconductivity since this would require a rather complex mathematical apparatus and is far beyond our main focus. Let us only acquire some superficial familiarity with vortices in superfluids and superconductors. In the next section we discuss the simplest vortices in superconductors that are mathematically described by the sine-Gordon equation.

We start with vortices in superfluids. Generally speaking, they are similar to vortices in the ideal liquid, but there is a dramatic difference — they are quantized.

---

[111]The theory of real superfluids was developed in 1953–1956. The physical picture based on the "rotonic" mechanism was made particularly clear by Richard Feynman (1918–1988).

[112]First the atoms of helium have no electric charge. Second, and more important, the electrons are *fermions* while helium atoms are *bosons*: two fermions cannot occupy the same quantum state (the Pauli exclusion principle) while any number of bosons may be in the same state (the Bose condensation phenomenon).

A more precise statement is the following. Consider a vortex in the superfluid helium which is schematically drawn in Figure 8.1.

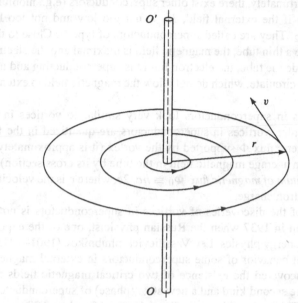

Figure 8.1 Superfluid vortex and its flow environment

Inside a very thin tube surrounding the vortex line OO′ the motion is vortical (nonzero vorticity, recall the end of the previous chapter). Outside the tube, the vorticity is zero and the dependence of the velocity of particles of liquid on their distance from the vortex line is described by the law $v = \kappa/2\pi r$. Here κ is quantized, i.e., $\kappa = nh/M_{He}$ where n is any integer number. This law is also valid for vortices in usual liquids but with arbitrary κ (not quantized). The fact that the "charge" (strength) of the vortices in superfluids is quantized is one of the consequences of the quantum nature of superfluids. This quantization was first introduced by Lars Onsager (1903–1976) in 1949 and further developed and used in the theory of superfluids by Feynman.

Note that quantizing the vortex is essentially quantizing the angular momenta of the helium atoms. Indeed, the two formulas above give the relation

$$(M_{He}v)r = nh/2\pi$$

where the left-hand side is the angular momentum of the atom and $h/2\pi \equiv \hbar$ is the quantum of the angular momentum.

Similar vortices may exist in some superconductors. As the Cooper pairs are charged, their motion generates electric currents and thus magnetic fields. It follows that inside the vortex tube there must exist a magnetic field. This means that an external magnetic field has to be applied to the superconductor. However, usual superconductors, such as aluminum, lead or tin, cannot support a magnetic field inside (this is called the Meissner effect). For this reason they do not support vortices. Fortunately, there exist other superconductors (e.g., niobium) which support vortices if the external field, H, is not too low and not too high, i.e., $H_{c1} < H < H_{c2}$. They are called superconductors of type II. Close to the vortex axis OO', inside a thin tube, the magnetic field is maximal and the electron liquid is normal. Outside the tube, the electron liquid is superconducting and undamped electric currents circulate, which do not allow the magnetic field to extend beyond the tube.

Thus, vortices in superconductors look very similar to vortices in superfluids.[113] In particular, vortices in superconductors are quantized in the following sense: the magnetic flux Φ supported by the vortex (it is approximately equal to the product of an average magnetic field in the tube by its cross section) is a multiple of the *quantum of magnetic flux* $\Phi_0 \equiv hc/2e$, where c is the velocity of light and e is the electron charge.

The history of the discoveries of vortices in superconductors is not straightforward. It began in 1937 when the Russian physicist, one of the experts in experimental low energy physics, Lev Vassilievich Shubnikov (1901–1945) discovered the unusual behavior of some superconductors in external magnetic fields. Actually, he discovered the existence of two critical magnetic fields for superconductors of the second kind and a new state (phase) of superconductors (Shubnikov's phase). This was not theoretically understood for almost 20 years. In 1952 L.D. Landau and V.L. Ginzburg advanced a general phenomenological theory of superconductors, applicable to a wide range of physical phenomena involving superconductors. Using this theory, A.A. Abrikosov theoretically discovered vortices and thus explained Shubnikov's experiments, suggesting that Shubnikov's phase is a state with vortices that actually form a periodic lattice. This result seemed to other physicists so strange that he could not publish it from 1952 until 1957, and even after 1957, this idea was accepted with difficulties and only after experimental proof of several predicted effects.

Nowadays, vortices in superconductors are used in technology. The biggest modern accelerators of particles use superconductors of type II working in the Shubnikov phase for their magnets. The theoretical ideas of superconducting vortices proved to be very useful in modern theories of fundamental interactions, which we review in the last section of this chapter. Before turning to these modern developments in fundamental physics, let us look at a simpler vortex existing in superconducting materials. This vortex can actually be described by the sine-Gordon equation and is a pure and beautiful physical realization of the sine-Gordon soliton.

---

[113] In fact, details of their structure are somewhat different and more complex but this is not important for us.

Josephson Solitons

The Josephson junction is essentially equivalent (mathematically) to the pendulum. To understand what it is, imagine that two pieces of a superconducting material have a contact at a point and that, at this point, they are separated by a very thin ($\sim 10^{-7}$ cm) layer (film) of an insulator (usually the oxide of the superconducting metal). Normal electrons cannot percolate through the insulating pellicle but the Cooper pairs are capable of filtering through it (physicists used to call this process superconducting *tunnelling*).

This phenomenon and extremely interesting and unexpected consequences of it were predicted in 1962 by a young English physicist, Brian Josephson. At that time he was a postgraduate at Cambridge University and the topic for his investigation was suggested to him by the American physicist Philip Anderson.[114] Josephson's short paper predicted a lot of physical effects that were soon checked in experiments.

For the physics community the Josephson effects were rather surprising. However, the superconducting tunnelling has a very interesting history. In 1932, German physicists W. Meissner and R. Holm found that the electric resistance of a small contact between two superconductors vanishes when the temperature is reduced below the critical point for both superconductors. It seems that this was the first encounter of physicists with one of the Josephson effects. However, nobody showed any interest in this effect. There was no theoretical explanation of it, and so it remained an isolated experimental puzzle, not unlike Russell's soliton. Possibly, experts thought that this result was a sort of experimental mistake which happens to even the best physicists.

One year later Meissner (with his collaborator R. Ochsenfeld) discovered the Meissner effect, and this discovery generated an avalanche of other experimental and theoretical studies in electrodynamics of superconductors. For describing the Meissner effect in mathematical terms, one has to modify the Maxwell equations. This was done in 1935 by brothers Fritz and Heinz London.[115] They supplemented the Maxwell equations by a new equation describing distribution of the magnetic field inside superconductors (London's equation). This equation contains a new characteristic of the superconductor – the London penetration depth, λ_L of the magnetic field.[116] For different superconductors λ_L may be slightly different but in general it is very small, $\lambda_L = 10^{-5} - 10^{-6}$ cm. London's equation is classical. However, it gradually became clear that superconductivity cannot be understood without quantum mechanics. Using the idea that superconducting electrons form something like a quantum liquid, F. London introduced, in 1950, the quantization of the magnetic flux in superconductors.

---

[114]He had a permanent position at Bell Laboratories and was visiting professor at Cambridge.

[115]They were born and started their scientific careers in Germany but moved to England when Nazis came to power; from 1939 F. London worked in the USA.

[116]I hope that the reader understands that the diameter of the vortex tube must not exceed λ_L. In fact it is of the order of λ_L.

Meissner did not forget about his work with Holm. His student I. Dietrich, in 1952, repeated their experiments and performed more detailed studies of the phenomenon for different temperatures. She explicitly demonstrated that the current flowing through the contact is superconducting. Unfortunately, theorists showed no interest in these new results and they were also forgotten.

Meanwhile, other effects of tunnelling were found. In 1957, a Japanese physicist, Leo Esaki discovered tunnelling in semiconductors and prepared the first semiconductor diode working on this effect. This made studying other tunnelling phenomena more fashionable. Thus, a Norwegian engineer, Ivar Giaver, who was working in the USA, decided to became a physicist. His first topic was the study of different tunnelling phenomena. In 1960 he observed in his experiments tunnelling of normal electrons between two superconductors separated by an insulator. His experiments were extremely important for verifying the theory of superconductivity. For us it is of interest that in some cases he probably saw the effect of Meissner and Holm, but he ascribed it to micro shunts in the insulating film (note that the Giaver effect can be seen for somewhat thicker insulating films than those used by Meissner and Holm).

Josephson did not know anything about the Meissner and Holm experiments. He simply solved a pure theoretical problem about tunnelling of the Cooper pairs through the insulating layer, derived formulas describing this phenomenon, and predicted other related physical effects. His consideration was not, strictly speaking, microscopic. A detailed microscopic theory of the Josephson effects was soon developed by Anderson and other theorists.

To really understand the Josephson effect, one must know quantum mechanics. For a simple qualitative understanding, it is sufficient to know that the Cooper pairs are described by a wave. The amplitude of this wave is proportional to the density of the Cooper pairs, while its phase (let us denote it by ϕ) describes interference phenomena. Let us suppose that the amplitude is the same in both superconductors. Then, only the phase is important. If two waves have the same phase, their sum is maximal. If the difference of the phases is π, the sum is minimal.

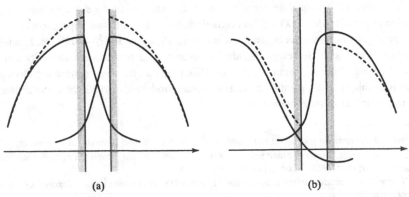

(a)   (b)

Figure 8.2

Let us see what happens with these two waves in a Josephson junction. In Figure 8.2a the waves in the left and in the right superconductors have the same phase on the border. After adding both waves, the resulting distribution of the superconducting electrons (Cooper pairs) remains symmetric, and thus no current will flow through the insulating film. In Figure 8.2b the phase difference between the two waves is $\pi/2$. Adding these waves does not remove the asymmetry between left and right superconductors and the Cooper pairs will flow through the insulating barrier to the left. This simple consideration shows that the current of the Cooper pairs (supercurrent), I, must depend on the phase difference $\phi \equiv \phi_1 - \phi_2$. Because nothing will change if we add a multiple of 2π to the phase difference, the current must be periodic in ϕ with the period 2π. The simplest guess is

$$I = I_c \sin \phi,$$

where the factor I_c is the critical (maximal) current. Josephson's quantum mechanical derivation gave precisely this formula. There may be additional terms but usually they are very small and do not change the main effect either qualitatively or quantitatively.

The second formula derived by Josephson is more difficult to motivate. It connects the potential difference between two superconductors and is proportional to the rate of change in the phase difference, $\dot{\phi} \equiv \Delta\phi/\Delta t$, namely;

$$V = (\Phi_0/2\pi c)\dot{\phi},$$

where Φ_0 is the magnetic flux quantum defined above and c is the velocity of light (in vacuum, of course). This formula shows that there is no potential difference for the time independent phase difference, and if we apply a constant external electric field, $V \neq 0$, the phase difference will be proportional to time t. According to the first formula of Josephson, the current will oscillate with high frequency depending on V. Thus the Josephson junction will emit electromagnetic waves (Josephson "generation"). There exists the inverse process — Josephson absorption. It follows that the Josephson junction may work as a generator or as a receiver of high frequency electromagnetic waves (in a frequency range inaccessible to other types of generators and receivers).

There are many other applications of the Josephson junctions — from physics to medicine. For us, most interesting is that in the extended Josephson junction, there may exist solitons that are similar to the Frenkel solitons. To understand this statement, consider the electric scheme of the Josephson junction, see Figure 8.3. The junction has a certain capacitance, C, and certain effective resistance, R, because the normal electrons are also tunnelling through the junction.

The current of the normal electrons in the resistance, I_N, is given by the Ohm law, $I_N = V/R$. The Maxwell displacement current, I_M, crossing the capacitance is $I_M = C\dot{V}$. The total current, I, flowing through the junction is the sum of I_N, I_M and of the Josephson current, $I_J = I_c \sin\phi$. Using the second Josephson equation, we can express I_N and I_M in terms of time derivatives of the phase

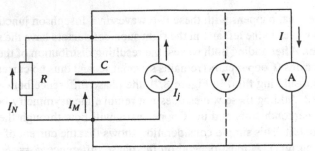

Figure 8.3 Electric scheme of the Josephson junction

difference ϕ and thus find the following equation for $\phi(t)$:

$$\ddot{\phi} + (RC)^{-1}\dot{\phi} + j_c \sin\phi = j,$$

where j_c and j are proportional to I_C and I, respectively. This equation essentially coincides with the pendulum equation on which two additional forces act: an "external force" j, and the "friction force" $-(RC)^{-1}\dot{\phi}$. Thus, whatever we know about the pendulum motions can be applied to the Josephson junction.

Now, it is not difficult to construct a chain of the Josephson junctions that is analogous to the chain of the pendulums or to the chain of atoms discussed in previous chapters. Its electrical scheme is presented in Figure 8.4, The inductance L, which is introduced when we electrically connect the junctions by wires, plays the role of the springs connecting pendulums or atoms. If we neglect the normal currents flowing through the junctions, we obtain the system equivalent to the Frenkel chain of atoms. Thus, in the continuum limit, the system is described by the sine-Gordon equation!

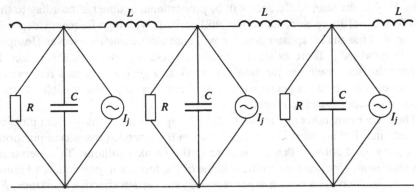

Figure 8.4 Electric scheme of a chain of the Josephson junctions

The physical system described by this equation is schematically depicted in Figure 8.5. It consists of two superconducting strips with the insulating film between them. Due to the Meissner effect, the magnetic field may exist only in the film and in adjacent thin layers of the superconductors, as shown for the case of one soliton (the arrows in the y-direction represent the magnetic field and the circles show the Josephson currents producing the magnetic field. The magnetic flux should be quantized. Accordingly, the one-soliton state carries one quantum of the magnetic flux (N solitons carry N quanta). Note that this is literally true if the junction is long enough compared to the characteristic length of the soliton, λ_J, which is called the Josephson length. This length is usually much larger than the London depth, λ_L, which is the penetration depth of the magnetic field into a superconductor.

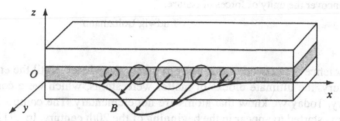

Figure 8.5 Soliton in a long Josephson junction

No electric field is shown in Figure 8.5. It may exist if the soliton is moving and is parallel to the z-direction. Moreover, if we apply an external electric potential, the soliton starts moving. Note that for the antisoliton, the magnetic field is parallel to $-y$ and under the action of the external potential, a soliton and an antisoliton will move in opposite directions. It must be clear that the normal tunnel currents give rise to dissipative forces (friction). Fortunately, they can be made very small and this makes it possible to observe the sine-Gordon solitons, their reflections from the boundaries of the long Josephson junction, their collisions with impurities (e.g., inhomogeneities in the insulating layer). It is not difficult to observe many soliton states, breathers, etc. A very useful feature of the Josephson solitons is that they are not difficult to operate by applying external electric voltage and current to the junction. By using artificially prepared inhomogeneities one can make bound states of solitons.

In this way one can store, transform and transmit information. In other words, long Josephson junctions may be used in computers. One of the most useful properties of such devices would be very high performance speed. Indeed, the characteristic time may be as small as 10^{-10} sec, while the size of the soliton may be less than 0.1mm. These ideas were seriously considered some time ago, but today they are not very popular, due to recent dramatic progress in semiconductor technology. Nevertheless, in perspective, the ideas of using Josephson solitons are of serious interest and should not be completely forgotten.

Note that other solitons may be of considerable interest for future computers. For example, vortices in type II superconductors might be used as memory cells. Magnetic solitons are now used in some memory devices. However, the time for the systematic use of solitons in computer applications is still ahead. So we will leave this matter and turn to some possible applications of solitons in the most difficult but most exciting research, which deals with the ultimate building blocks of physical matter.

Elementary Particles and Solitons

The main goal of the natural sciences is
to uncover the unity of forces of Nature.

Ludwig Boltzmann

Modern science is uncovering this unity at a very deep level. Until the end of the 19th century, the ultimate blocks of matter were atoms, which were considered elementary. Today we know that atoms are not elementary. The composite models of atoms started to appear in the beginning of the 20th century. In 1911, Ernest Rutherford proposed the planetary model of atoms (the central part of any atom is a positively charged atomic nucleus and the p negatively charged electrons play the role of planets). Two years later Niels Bohr quantized the Rutherford atom. Only 65 years ago did it become clear that the nuclei consist of protons and neutrons, which for some time were considered elementary. More than 30 years later physicists found that the protons and neutrons, as well as many other particles discovered in cosmic rays and in experiments on accelerators of protons, are not elementary, and their constituents, *quarks*, were introduced by Murray Gell-Mann and George Zweig in 1964.

Now we know that the building blocks of matter in our universe are *leptons* (electron, muon, τ-lepton, electronic, muonic and τ neutrino) and 18 quarks (six different types called "flavors," each flavor is realized in three "colors"). All these particles are fermions and have intrinsic angular momentum (spin) one half. Note that the only known stable particles are: electrons, protons, photons, and neutrinos. All other particles are unstable: the neutron has the longest the lifetime — approximately 15 minutes;[117] other observable particles decay for less than 10^{-5} seconds. Quarks have fractional electric charges (in units of the elementary charge e) and are not directly observable; we may regard them as peculiar quasiparticles that exist inside *hadrons*, which are bound states of quarks (e.g., protons and neutrons contain three quarks, mesons are bound quark–antiquark pairs). Quarks are permanently confined in hadrons but, possibly, they were free at early stages of the formation of the universe. Although many photons travel across our universe, they are not usually considered as building blocks of matter — they are rather carriers of electromagnetic interaction between particles. Other bosons are carriers of other interactions.

---

[117]In nuclei the neutron may be stable because the decay of the bound neutron may be not allowed by energy conservation.

Boltzmann knew of only two fundamental forces of Nature: gravitational and electromagnetic.[118] In addition to the electromagnetic forces, which bound electrons and nuclei into atoms, we know of two more types of forces — strong and weak. *Strong* forces bind three quarks into baryons (protons, neutrons, etc.) and quarks and antiquarks into mesons. Protons and neutrons are kept bound in nuclei by somewhat weaker strong forces. Strong forces are also responsible for decays of hadrons (i.e., strongly interacting particles — baryons and mesons). *Weak* forces are responsible for relatively slow decays of particles – the decay of the neutron into the proton, electron and neutrino or decay of the muon into the electron and two neutrinos, etc. As distinct from the gravitational and electromagnetic forces, the weak and strong forces have short ranges. The range of strong forces (the average size of hadrons) is approximately 10^{-13}cm while the range of the weak forces is, on average, 10^{-16}cm. For comparison, recall that the typical size of an atom is 10^{-8}cm.

This distinction is very important, so let us explain it in more precise terms. The Coulomb (or gravitational) forces acting between electrically charged particles (or massive bodies) are inversely proportional to the distance separating them. This means that in quantum theory, the exchanged particles (photons or gravitons) are massless. The forces generated by the exchange of massive particles are decreasing very fast for distances exceeding the Compton wavelength of the exchanged particle, $\lambda \equiv \hbar/mc$. In mathematical terms, strong and weak potentials depend on the distance r like $g \exp(-r/\lambda)/r$ (strictly speaking, the effective coupling "constant" g may also depend on r, see the discussion below). This means that the particles responsible for the strong forces must be about 200 times heavier than the electron, while those responsible for the weak interactions must be about 100 times heavier than the proton and 200,000 times heavier than the electron. This looks rather obvious today, but it was not at all clear 65 years ago, when strong and weak interactions were introduced for the first time to explain new experimental facts.

Strong forces were first introduced by Heisenberg in 1932 to explain binding of protons and neutrons into nuclei. In particular, Heisenberg considered the bound state of the neutron and proton — the deuteron. Soon, he and other scientists (D.D. Iwanenko, I.E. Tamm, A. Nordsiek) tried to relate these forces to the theory of weak interactions advanced by Fermi in 1933 (at that time it was called the theory of β-decay). This attempt was not successful and a young Japanese physicist Hideki Yukawa clearly understood this. As he recalled later "... I thought: let me not look for the particle that belongs to the nuclear force among the known particles.... The crucial point came to me in October (1934). The nuclear force is effective at extremely small distances. My new insight was that this distance and

---

[118]The electric and magnetic forces were the only fundamental forces of Nature, which seemed to be relevant to the small scale structure of matter. The gravitational forces are only important in large scale phenomena, and thus may be completely neglected when we study the atomic structure of matter. Indeed, the gravitational attraction between the electron and the proton is approximately $2 \cdot 10^{39}$ times less than the electric attraction.

the mass of the new particle are inversely proportional." Thus Yukawa introduced the new particle, now called the π-meson, or the *pion*. Although it was experimentally discovered only in 1947, Yukawa's idea was enthusiastically accepted, and strongly influenced the development of the quantum theory of fields and of new physics in general before the mid-forties.

The development of the quantum field theory started with quantizing Maxwell's theory of the electromagnetic field. The quantized theory permitted a description of the processes of creation and annihilation of photons — the quanta of the electromagnetic field, which have zero mass and spin one. This completed the concept of the wave particle duality. Constructing the theory of the electron field and its coupling to the electromagnetic field (Quantum Electrodynamics or QED) was a highly nontrivial thing and required more than 30 years of efforts among the best theoretical minds in the world. It started in 1926–1927 with two fundamental papers by Dirac. From 1927 to 1932 the main concepts and methods of QED were developed, and the main difficulties were revealed (by Dirac, Born, Heisenberg, Jordan, Klein, Pauli, Fock and others). The principal difficulty was that attempts to calculate physical effects with high precision invariably led to infinities (divergent integrals). Only first approximations gave sensible results. The origin of this difficulty was, apparently, in the point-like nature of the particles. Some theorists tried to supply particles with internal structure and thus avoid divergences. However, this introduced a high degree of arbitrariness into the theory and was not fruitful.

The successful solution of these problems was found soon after World War II. The solution was the so-called renormalization theory, which allowed computation of all physically measurable effects by consistent isolation of the infinities. The final results were expressed in terms of only measurable physical constants and variables. So quantum electrodynamics became a most precise theory. However, it was clear that this theory is not the final physical theory of the world. First, the origin of the electromagnetic coupling and its small coupling constant[119] has no explanation in the framework of QED. The mass of the electron is also a free parameter of the theory. There is no explanation why the muon, having the same electromagnetic properties as the electron, is 200 times heavier. Finally, electromagnetic properties of strongly interacting particles (hadrons) cannot be described by QED with a reasonable precision. In other words, QED is an incomplete theory.

Naturally, many theorists attempted to apply the ideas and methods of QED to weak, strong and gravitational interactions. These attempts were unsuccessful for a long time, the reasons of the failure being different for the strong and the weak interactions. Strong coupling constants are so large that perturbation theory is unapplicable. For weak and gravitational interactions the coupling constants are

---

[119]The dimensionless measure of the electric force is the fine structure constant $\alpha \approx 1/137$, which is proportional to the square of the elementary electric charge e divided by $\hbar c$. The dimensionless factor depends on the choice of units. We use here the units in which $\alpha = e^2/4\pi\hbar c$ and the Coulomb potential for two electrons is $V = e^2/4\pi r$.

very small but the renormalization theory does not work and infinities cannot be removed. Different attempts to overcome these problems invariably failed until the 1970s when the consistent theories of weak, electromagnetic and strong interactions were finally constructed. The key for this achievement was found in 1954 by Chen Ning Yang and Robert Mills. They proposed a nontrivial generalization of Maxwell's theory, in which interaction is transmitted by several massless "photons" which may bear charges.[120] This theory is renormalizable and mathematically very beautiful. However, at first, it looked useless for both the weak and strong interactions because the charged "photons" were massless and thus could not generate short range forces. If you simply add to the theory terms supplying masses to these particles, the theory is not renormalizable.

Before any consistent application of Yang–Mills theory to the real strong and weak interactions became possible, several important ideas had to be introduced. In 1957 Schwinger proposed replacing Fermi's theory of weak interactions with a unified theory of weak and electromagnetic interactions, in which weak interactions are mediated by charged massive vector (spin one) mesons.[121]

In 1958–1964 several useful ideas were proposed for intermediate bosons and unification. In particular, symmetries of strong and weak interactions were uncovered. This was one of the main ingredients of the future construction of the complete theory of strong, weak and electromagnetic interactions. Although at that time quantum field theory was rather unpopular, several field theory models were proposed and studied. Their common deficiency was nonrenormalizability. In fact, this was one of the main reasons for abandoning quantum field theory by the theoretical community.[122]

The year 1964 was extremely fruitful for particle theory although it became clear only in retrospective. Three quarks were invented and, moreover, from purely aesthetic consideration, several people added a fourth ("charmed") quark. At that time, the charmed quark was not necessary for strong interactions but later proved to be crucial for constructing the renormalizable electroweak theory. The second important idea appeared in the paper of Peter Higgs who proposed that the "photons" in Yang–Mills theory could acquire masses by a mechanism similar to the known mechanism responsible for the Meissner effect in superconductors. Indeed, inside a superconductor photons become massive due to coupling with Cooper pairs, and the London penetration depth is nothing but the Compton wavelength of these "massive photons." Instead of Cooper pairs, Higgs introduced a scalar field (describing particles with zero spin, the "Higgs particles") and arranged coupling

---

[120]We call these particles "photons" because, like real photons, they have mass zero and spin one (the corresponding fields are vector fields similar to electric and magnetic fields).

[121]These particles are called intermediate bosons and were first discussed by Yukawa. The first attempt to unify electromagnetic and weak interactions was made in 1938 by O. Klein. And these ideas were forgotten for many years.

[122]For example, when N. Cabibbo proposed (in 1963) the most important generalization of Fermi's theory, he practically did not use field theoretical concepts in his paper.

in such a way that the configuration of fields with minimal energy corresponds to massive Yang–Mills bosons (in fact, some of them may remain massless).

Thus the stage was ready for the appearance of the electroweak (EW) theory. It was proposed in 1967,[123] and its final version was formulated and proved to be renormalizable in 1971. After this extremely important theoretical achievement, the electroweak theory became rather popular and its experimental consequences were worked out. Although the masses of the intermediate bosons are very large (about 100 of the proton masses) and their lifetime very small, less than 10^{-24} seconds, they were observed in experiments with high energy accelerators in Fermilab (near Chicago), CERN (European Center of Nuclear Research near Geneva) and SLAC (Stanford Linear Accelerator Center). We need not (and cannot) go into a detailed description of this beautiful theory and the ingenious experiments that proved it. Note only that the highly nontrivial proof of renormalizability of the electroweak interactions was given by a young Dutch theorist Gerardus 't Hooft, (also a recent Nobel Prize winner) who somewhat later found that electroweak theory predicts a very interesting soliton. We will soon return to this idea and now briefly describe how the Yang–Mills fields were found to give the fundamental theory of strong interactions.

After the invention of quarks, it soon became clear that all known mesons (i.e., hadrons with integer spins and zero baryon charge) are composed of a quark and an antiquark, while all known baryons (i.e., hadrons with half integer spins and baryon charges +1 or −1) are composed of three quarks. To have correct quantum numbers for baryons, theorists were forced to introduce a new quantum number. It was called "color," because each quark exists in three different states (of different "colors"), while the hadrons are "colorless" ("white"). To describe forces acting between colored quarks, the colored "photons" (massless vector mesons) were introduced. The next step was to treat them as Yang–Mills bosons. This theory, which is called quantum chromodynamics (QCD), is renormalizable and, in addition, has a very important property — the behavior of the forces between quarks is much better than in QED. In fact, quarks that are close enough to each other behave as if free; this property is called "asymptotic freedom."

In more precise terms, the effective charge in QCD and other gauge theories (Yang–Mills theories and QED) depends on the distances between quarks. Effective charges behave somewhat differently for electromagnetic, weak and strong interactions. Effective electromagnetic coupling grows at small distances, while weak and strong couplings decrease. The most interesting thing is that at very small distances they may become equal! These distances are not only small but fantastically small, about 10^{-30} cm. It is impossible to directly check the theory at such small distances. But the fact that all interactions become equally strong dramatically changes our understanding of the structure and evolution of our uni-

---

[123] We are not attempting here to give a history of the development of this theory. Note only that Steven Weinberg, Abdus Salam and Sheldon Glashow became Nobel Prize winners for their outstanding contribution to creating EW theory.

verse. Indeed, we now believe that at such distances, weak, electromagnetic and strong interactions are unified and there is no principal distinction between quarks, hadrons and leptons (this is called the Grand Unification). In particular, this means that the proton may decay into leptons. Although the lifetime of the proton is much longer than the age of the universe, the existence of decay is of utmost importance for the fate of the universe.[124] Another intriguing property of the Grand Unification Theory (GUT) is that it probably must unify fermions and bosons by the so-called supersymmetry (Supersymmetric Grand Unification, or SUSY GUT). The theories of Grand Unification and supersymmetry predict many new particles and interesting new relations in the world among the presently known particles. Some attempts are being made to check these predictions and search for new particles at existing accelerators, for example, at the new giant accelerator in Geneva. Up to now the search for new particles and for an understanding of effects that cannot be explained without GUT have failed to give positive results.

Almost all theorists believe that Grand Unification exists but a detailed theory has not yet been created. It may be a (supersymmetric) gauge theory or a more general field theory, the quanta of which are not point particles but some extended objects. These may be one-dimensional strings, two-dimensional membranes, etc. It may happen that even field theory will be abandoned and the most fundamental notions of space and time will be changed. There are several arguments supporting the idea that the final theory of matter most probably will not be a field theory of point particles (or, a theory of local fields). First, we have no consistent quantum theory of gravity and thus we cannot unify gravitational interaction with electroweak and strong interactions. Without this unification we have no hope of explaining the origin of our universe as well as its fundamental properties today and its future fate.

There are also more practical problems in particle physics and cosmology that must be solved before quantum gravity can be created and unified with other interactions. We will not go into details. Note only the most fundamental fact. The quantum effects of gravity must have been most important when the universe was as small as the Planck length,[125] $l_{Pl} = (\hbar G/c^3)^{1/2} \approx 10^{-33}$ cm. This unimaginably small dimension is constructed of the Newton gravitational constant G and is a characteristic length of gravitational interaction. For this dimension of the universe, gravitational forces become strong and cannot be neglected anymore. Possibly, there do not exist objects of dimension less than the Planck length and thus a theory including gravity must not be a theory of point particles (or local fields). It is thus probable that the ultimate building blocks of matter are some extended objects. Even if they are not solitons, they may somehow be related to

---

[124]In 1967 Andrei Dmitrievich Sakharov demonstrated that small instability in the proton is necessary to explain the present state of the universe in terms of the current Big Bang model. At that time the idea of proton decay seemed absolutely crazy to particle physicists.

[125]This is the Compton wavelength of the Planck mass, $M_{Pl} = (\hbar c/G)^{1/2} = 1.3 \cdot 10^{19} m_p \approx 2 \cdot 10^{-5}$ g, where m_p is the proton mass. The intermediate bosons in GUT models would be 1000 times lighter.

solitons. We will very briefly describe these hypothetical solitons after a brief review of less fundamental but, unfortunately, also hypothetical solitons.

Yang–Mills theories of weak and strong interactions, as well as GUT, predict diverse solitons. Some of them are more or less familiar to us, others are quite new and unusual. The fate of our universe may depend on a beautiful soliton which is predicted by unified theories. In addition to other puzzling properties, this soliton bears a magnetic charge. For this reason, it is called the *magnetic monopole*.

It is well known that the Maxwell theory does not allow for the existence of isolated magnetic charges.[126] In this sense, there is no symmetry between electric and magnetic fields. With time, this state of affairs acquired the status of a dogma and it was seriously shaken only in 1931 by Dirac. He reanalyzed the problem of the magnetic charge and found that, in the framework of quantum mechanics, isolated magnetic charges may exist. Most interesting was the fact that the existence of a single magnetic charge in the world would result in quantizing the electric charge! This follows from the famous Dirac formula for the magnetic charge

$$g = nhc/e \equiv 2n\Phi_0,$$

where n is an integer and Φ_0 is the magnetic flux quantum (note that the last quantity was introduced twenty years later!). One can see that the symmetry between the electric and magnetic forces is not completely restored, because the magnetic charge g is inversely proportional to the elementary electric charge e. This sort of symmetry relations is now called *duality*, which is one of the most popular words in the theoretical community.

While the elementary electric charge is small enough to use conceptually simple methods (perturbation theory) for computing quantum effects, the magnetic charge is so large that there exists no computational method for dealing with quantum effects of magnetic charges. In other words, quantum electrodynamics is almost linear theory — in first approximation, the interaction between electromagnetic field and electrons may be neglected. This is absolutely impossible for the magnetic charges — coupling them to the electromagnetic field is very strong and there exists no linear approximation. In fact, we do not know any sensible approach to quantizing the theory with magnetic charges. It may even happen that, in quantum theory, free magnetic charges (magnetic monopoles) cannot exist and there may exist only monopole-antimonopole pairs connected by a thin string with a high magnetic field inside.

Such considerations led Julian Schwinger to propose that quarks are magnetically charged. This hypothesis was suggested in the pre-QCD era and now is not considered seriously in its original form. However, generalizations of these ideas are still popular and useful. The picture of the hadron as made of quarks connected by thin strings with a strong chromoelectric (or chromomagnetic) field inside still seems attractive, in spite of the fact that nobody can prove it by real computations.

---

[126]The ends of a long and thin magnet partly imitate magnetic charges of opposite signs but one can never isolate these "charges."

Certainly, this picture is oversimplified and the real hadrons are much more complicated. At best, this picture may serve as a first approximation for describing a world in which free quarks do not exist due to strong coupling between them. It is certainly more than a freak of the imagination but less than a theory.

Experimental search for the Dirac monopole was not successful. Apparently, Dirac monopoles, if they exist, are extremely rare. While people gradually lost interest in monopoles, a beautiful reincarnation of the Dirac monopole occurred in 1974. Young theorists Alexander Polyakov from Moscow and Gerardus 't Hooft, then working at CERN, independently discovered that in electroweak theories there may exist a new topological soliton bearing the minimal magnetic charge. Although it is an extended object, its magnetic field at large distances is similar to the magnetic field of the point-like Dirac monopole.

Both theorists used a deep analogy between the vacuum of gauge theories and superconductors. In this analogy, the role of the field of the Cooper pairs was played by the Higgs field. As we mentioned above, in gauge theories there are analogs of the electric and magnetic fields. However, unlike QED, which is abelian gauge theory (having abelian symmetry), the EW theory has a more complex symmetry (non-abelian) and belongs to a wider class of non-abelian gauge theories.[127] As distinct from QED, gauge bosons in non-abelian gauge theories have nonlinear couplings between them. As a result, in these theories there may exist stable solitons.

Polyakov–'t Hooft monopoles have most interesting properties in Grand Unified theories. They are very small and very massive objects. Their masses may be as large as 1/1000 of the Planck mass while their dimensions may be as small as 1000 of the Planck length. They must have a complex internal structure. In the core of the monopole, weak, electromagmetic and strong forces are equally strong, and for this reason they may be very dangerous for protons. In fact, when they meet protons they may catalyze their decay. This dramatic consequence of monopole existence was discovered in 1981 by a young Russian, Valerii Rubakov, and an American, Curtis Callan. Although the GUT predicts proton instability, its lifetime in the present state of the universe highly exceeds the lifetime of the universe. However, if there were many monopoles in the universe, its fate would be very sad. We are very fortunate that monopoles are so rare nowadays, but this fact has to be explained by cosmological theories. The monopole triggered new ideas in cosmology and our concepts of the initial stages of the universe are now rather different from those in the pre-monopole era.

---

[127]Pure gauge theories, not coupled to matter fields, may be classified by their symmetries. One then tries to add fermion matter fields in a way that maximally preserves the symmetry. However, in any realistic gauge theory, except QED, the original gauge symmetry is broken. Different patterns of gauge symmetry breaking are called different *phases* of the gauge theory. We know the Coulomb phase (QED), the Higgs phase (EW), and the confinement phase (QCD). There exists a more refined classification of different phases in gauge theories. For us, it is only important to understand that the vacuum in gauge theories is similar to the vacua in condensed matter physics (superfluidity, superconductivity, ferromagnetism, etc.).

The Grand Unified theories teach us that there may exist other extended objects that played an important role in the early evolution of the universe. Using the analogy of the gauge theory vacua to superconductors and magnetics, the reader may guess that these objects may be flux tubes (called cosmic strings) and domain walls. Possibly, they were as abundant as the monopoles at the very early stages of the universe's evolution, but every search for these objects in the now observable universe has given up to now negative results. Nevertheless, they are of great conceptual importance, because they relate the processes occurring at the largest and smallest scales in the Universe. Now it is clear that understanding the large scale construction of the universe is not possible without knowledge of its smallest building blocks, and vice versa. This idea is presented in Glashow's picture of the universe, Figure 8.6.

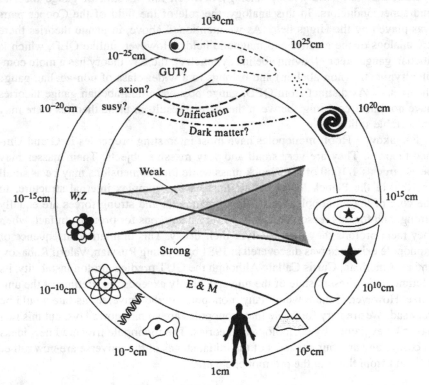

Figure 8.6 Picture of the universe

Before we briefly review soliton ideas related to the microworld, let us have a look at a very schematic picture of the small scale structure of the world. We show blocks of matter ("particles"), their interactions and transmitters of these interactions ("mesons"). We also give approximate values of masses of the mesons and their Compton wavelengths, which are their characteristic radii (according to Yukawa). Note that the experimentally verified part of this scheme lies above

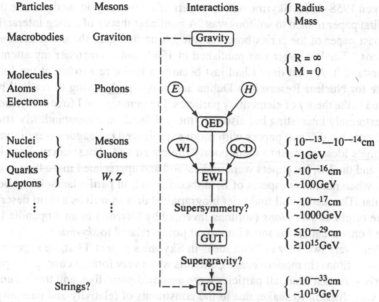

Figure 8.7 A Schematic Picture of the Microworld: Particles and Forces.
E and H–electric and magnetic forces. QED–quantum electrodynamics.
WI–weak interactions. QCD–quantum chromodynamics,
the theory of strong interactions. EWI–electroweak interactions.
GUT–grand unified theories of weak, electromagnetic and strong interactions.
TOE–hypothetical "theories of everything"
unifying all forces of Nature, including gravity.

"Supersymmetry". Particles, predicted by the supersymmetry (supersymmetric partners of ordinary particles) have not yet been discovered. "Supergravity" is a supersymmetric generalization of the Einstein theory of gravity. This theory has not yet been completely developed. "Theories of everything" are very speculative attempts to unify all interactions, including gravity. Note that, at this level, a distinction between particles (matter) and mediators of interactions (mesons) becomes meaningless and most probably the ultimate building blocks of the world must be some new structures. These structures may be superstrings and the main candidates for such a unification are superstring theories. Different string theories are now believed to be interconnected in a frame of a more general "M-theory." These theories use all known mathematical ideas and tools and require developing new mathematical ones. Here we only mention a few ideas about possible relevance of soliton theory to modern theories on particles and strings.

Solitons emerged in particle physics at the end of the 1950s and the beginning of the 1960s, even before Kruskal's and Zabusky's work, and quite independently of the devolopment described in the previous chapters of this book. They were introduced by a very original theorist, Tony Skyrme, in a series of papers published between 1958–1962. Skyrme was an expert in nuclear physics and particle theory. His first paper related to solitons was "A nonlinear theory of strong interactions". The last paper of the series bore the title "A unified field theory of mesons and baryons." The last paper was published in 1962, and I remember my attempts to understand it. At that time I had just became a junior researcher at the Joint Institute for Nuclear Research in Dubna and was eagerly trying to read all papers related to the theory of elementary particles. I remember that I found Skyrme's paper extremely interesting but also extremely difficult to understand. My attempts to discuss his difficult papers with more experienced colleagues were in vain — Skyrme's ideas seemed to them poorly motivated and almost crazy. Both in Europe and the US, his papers were understood and appreciated more than a decade later, when nonlinear aspects of field theories and, in particular, solitons became popular. The reader will find brief biographical data as well as a short description of the origin of *skyrmions* (solitons invented by Skyrme) in an Appendix below. Here I only mention the most important points related to *skyrmions*.

There were two very difficult ideas in Skyrme's papers. First, he regarded protons as solitons (in modern usage), and this was a very foreign concept for particle theorists. At that time, all particles were assumed point-like, and this assumption was very difficult to shake, due to the constraints of relativity and causality. The effective structure of the proton was believed to be generated by its interaction with pions and other particles, which, so to speak, dressed the proton with a cloud of virtual mesons (living for no more than 10^{-23} seconds). Another, even more severe difficulty was that skyrmions, which were fermions, were constructed from Bose fields. Thus, the proton was considered to be built of pions, which was in contradiction to all established theoretical concepts.[128] To explain the ideas of making fermion particles from nonlinear bosonic fields, Skyrme analyzed in detail a simpler two dimensional sine-Gordon field theory. He derived several exact solutions of the sine-Gordon equation independently of previous work which had been unknown to him. With this simple example he demonstrated that his solitons behave like fermions. Rigorous proofs of the equivalence between the sine-Gordon theory and the theory of fermions were given 15 years later by others.

An additional difficulty for understanding the skyrmion was presented by the topological nature of its stability. According to Skyrme, the baryonic charge, conservation of which is so fundamental for our existence, is a topological charge, analogous to the well-known to us "charge" of dislocations but much more difficult to define because the skyrmion is a three-dimensional object. At that time topology was not among standard tools used by theorists. After invention of the monopole they gradually became more familiar and now topological concepts and methods belong to the standard equipment of any theorist. This became

---

[128]At that time the distinction between fermions and bosons was akin to a dogma. It was possible to imagine that mesons are made of baryons and antibaryons, like nucleons in the Fermi–Yang model, but the reverse looked crazy. Note that the first paper on supersymmetry had only been published in 1971 and the ideas of a symmetry between fermions and bosons were more or less generally accepted only at the end of the 1970s.

possible due to the joint efforts of mathematicians, (M. Atiyah, S. P. Novikov, A.S. Schwarz) and theoretical physicists (A.M. Polyakov, E. Witten, B. Zumino). One may say that, as we probe the microworld on deeper and deeper levels, the collaboration between mathematics and theoretical physcists becomes stronger and stronger. Topological ideas were further developed in many papers devoted to the gauge theory monopoles.

On the other hand, intuitive physical ideas among the great physicists of the 19th century continue to influence modern science. Skyrme himself admitted that he was influenced by Kelvin's ideas on vortex atoms. An even more direct modern realization of Kelvin's vortex atoms was proposed in 1970 (100 years after publication of Kelvin's paper on vortex atoms) by the Russian mathematician Ludwig Dmitrievich Faddeev, who made important contributions to nuclear physics, gauge theory and soliton theory. He proposed a nonlinear relativistic field theory, in which knotted vortices could be stable.[129] However, it was not possible to solve this theory either analytically or numerically, and thus the existence of stable knotted vortices remains a plausible conjecture supported only by qualitative arguments. Nineteen years later, Faddeev and A. Niemi in the paper "Stable knot-like structures in classical field theory",[130] announced a numerical proof of the existence and stability of knotted solitons. They may be relevant or irrelevant to particle theory, because the model is classical and it is not clear how to quantize it. However, these knotted solitons are of great interest for many branches of science — from cosmology to polymer physics and molecular biology. The knotted solitons are essentially knotted tubes. If the diameter of the tube is very small in comparison to the characteristic dimensions of the soliton, we might regard the soliton as a knotted string. This brings us to our last topic — the theory of fundamental strings.

Strings as mediators of interaction were first introduced in the beginning of the 20th century by J.J. Thomson. His idea was the following. The electric and magnetic force lines connecting fast moving electrons form a thin tube; outside the tube the fields vanish. This is analogous to the modern QCD string connecting a quark and an antiquark. However, there is no such solution in Maxwell–Lorentz electromagnetic theory, which was only known to J.J. Thomson. For this reason Thomson and his followers failed to prove this conjecture, Thomson's idea was completely forgotten, and the strings in QCD were introduced by analogy with the flux tubes in superconductors. The Thomson picture in QCD may be realized if the vacuum is in the confining phase. Then a quark and an antiquark at rest are connected by a tube with the chromoelectric (chromomagnetic) field inside. The diameter of the tube must be of the order of the hadron dimension, i.e., 10^{-13} cm. This diameter is defined by the properties of the vacuum and is independent of the distance between the quarks. Thus the interaction energy is proportional to

---

[129] A simple ring is not stable; it either shrinks to a point or blows up. If it is knotted, some parts of it repel while other parts attract each other. Thus, the knotted ring may become stable.

[130] See *Nature*, vol 387, 1 May 1997.

distance, and the attractive force is independent of distance. This is what we call the confinement of quarks inside mesons. The tubes connecting three quarks inside baryons must be somewhat different, but this is of no concern to us. If the quark–antiquark tube is split into two new tubes, at the new ends appear a new quark and antiquark, so that one meson is broken into two new mesons (compare this to breaking a magnet!).

Now, suppose that the tube is very thin and that it has only two possible deformations — movements in the transverse direction and lengthening. Then one can find the spectrum of all mesons, which is very simple and agrees with the experimental one: the squared masses of the mesons, having the same quantum numbers (charge, strangeness, isospin), are proportional to their spins. The dependence is roughly universal, i.e., the slopes of the straight lines are roughly the same for different families of mesons. Here I do not follow the historic development of the ideas. These straight lines, called the Regge trajectories, were first discovered experimentally and the string picture emerged much later, as the result of evolution of apparently very different ideas. We will not go into detail because neither the tube nor the string pictures have been rigorously derived in the framework of QCD. Moreover, the string picture (infinitesimally thin, one-dimensional strings) was shown to be inconsistent in our three-dimensional space.

The reason was that the string should be relativistic and quantum, i.e., it must satisfy requirements of special relativity theory and quantum mechanics. Both requirements happen to be consistent only in 25-dimensional space (26-dimensional space-time). In trying to escape this problem, theorists invented supersymmetric strings (in fact this was the first application of supersymmetry) but such strings might exist in 10-dimensional space-time. So it became clear that relativistic strings have no relation to hadrons and most theorists stopped thinking about them. However, a few theorists continue to work with these strange objects and soon they found a completely new interpretation of string theory.

One of the most puzzling properties of strings was the following. The strings representing mesons or baryons must be "open," having quarks on their ends. However, the complete theory must also incorporate closed strings. The simplest one, topologically equivalent to a ring, was shown to have spin two and zero mass. Thus there must exist a massless particle with spin two. However, the existence of such a hadron would contradict all that we know about hadron physics. What does this mean? Two young theorists from CALTECH (Pasadena, Ca), John Schwarz and Joël Scherk, were very much puzzled by this fact. More than ten years later, when the new interpretation of superstrings became very popular, John Shwarz recollected:

> In 1974–75 I worked with Joël Scherk at CALTECH. We were struck by the fact that string theories had never yielded to our numerous attempts to shift the masses to other values. In particular... in the closed string sector inevitably appeared a massless spin-two state. At some point it occurred to us (I don't remember whether it was Joël or I who said first) that maybe the massless two-spin state is a graviton. This innocent remark had profound conse-

quences: It meant that we were no longer discussing a theory of hadrons. It meant that the natural length scale for strings is 10^{-33}cm (the Planck scale) rather than 10^{-13}cm... Most important, it meant that we have a candidate for a quantum theory of gravity! Once we realized we were talking about gravity, our attitude toward the extra dimensions also changed. We realized it would be perfectly reasonable to take them seriously as real and physical (as demanded by the theory), with a Kaluza–Klein interpretation.

This was a resurrection of the rather old idea of unification of gravitational and electromagnetic interaction. In 1919, a German physicist Theodor Kaluza (1885–1952) proposed treating the electromagnetic field as an additional gravitational potential in the five-dimensional world. His theory was rather formal, because the only effect of the fifth dimension was the appearance of the electromagnetic potential (there must be no dependemce of the physical variables on the fifth coordinate). For this reason, Kaluza's theory did not attract much attention from leading physicists. Nevertheless, seven years later, several papers developing Kaluza's ideas were published. A Russian physicist Georgii Aleksandrovich Mandel rediscovered Kaluza's theory and treated gravity in five dimensions in more detail. Following Mandel, V.A. Fock tried to work out some physical consequences of the fifth coordinate. Very similar results were independently obtained by O. Klein.[131]

Unfortunately, in all the beautiful papers mentioned on Kaluza's theory, the physical meaning of the fifth coordinate remained unclear. Only in 1938 did A. Einstein and his young collaborator P. Bergmann arrive at a modern understanding of Kaluza's theory. They proposed that the fifth dimension is physical but its effects are difficult to observe because the fifth dimension is compactified into a very small circle. To understand this idea, imagine an infinitely long cylinder having a very small radius (in comparison to our macroscopic dimensions). If you move on its surface in the direction perpendicular to its axis, you soon return to the starting point. This is the compact coordinate. The other coordinate, along the axis, is the normal one (noncompact). In modern theories using Kaluza's idea there are several compact coordinates and, of course, four normal coordinates describing our space-time. Gravitational potentials related to the compact coordinates generate gauge (Yang–Mills) fields.[132]

In this way one may hope to obtain a unified theory of all interactions, including gravity. Then the compactification radius (which also is the characteristic length of the string) must be of the order of the Planck length, l_{Pl} (and less than $10^3 l_{Pl}$). The corresponding characteristic mass is of the order of the Planck mass, M_{Pl}

---

[131] The commonly used term "Kaluza–Klein theories" thus seems to be historically unjustified. It would be better to call the theories with extra dimensions interpreted as some fields either "Kaluza theories" or "Kaluza– Mandel–Klein–Fock theories."

[132] The many charges required by electric, strong, and weak interactions may, possibly, be interpreted as topological charges of some solitons living on strings. For example, a closed string on a cylinder bears a topological charge depending on the number of unshrinkable loops on the string.

(and larger than $10^{-3} M_{Pl}$). It must be clear that all particles that were observed and willl be observed in accelerator experiments are massless on this scale. The origin of their masses and couplings is one of the many problems that have to be solved by string theorists. Experiments will not help them in this pursuit, because no experiments may directly probe the structure of the world on the distances of the order of l_{Pl}. They may only use the intuition developed by the preceding generations of physicists as well as purely mathematical structures. The soliton ideas, which are both intuitively clear and mathematically deep, may prove very useful even on this border of our knowledge.

Above, I tried to give you an intuitive picture of solitons. I did not even touch mathematical theories behind this simple picture. To fully appreciate the beauty of the theory of solitons as well as to succesfully apply it to solitonic phenomena in different domains of science, one should try to study the necessary mathematics (some hints and advice are given in the Mathematical Appendix). I hope that some young readers of this book will try this intellectual adventure. For them, I finish this book with three quotations from some great physicists of the 20th century. They were speaking about physics but what they said is relevant to all sciences that are mature enough to use modern mathematics.

> It seems to be one of the fundamental features of nature that fundamental physics laws are described in terms of great beauty and power.
> As time goes on, it becomes increasingly evident that the rules that the mathematician finds interesting are the same as those that nature has chosen.
> Paul Dirac

> The enormous usefulness of mathematics in the natural sciences is something bordering on the mysterious and there is no rational explanation for it. It is not at all natural that "laws of nature" exist, much less that man is able to discover them. The miracle of the appropriateness of the language of mathematics for the formulation of the laws of physics is a wonderful gift which we neither understand nor deserve.
> Eugene Wigner

Appendix I
Lord Kelvin *On Ship Waves*

"WAVES" is a very comprehensive word. It comprehends waves of water, waves of light, waves of sound, and waves of solid matter such as are experienced in earthquakes. It also comprehends much more than these. "Waves" may be defined generally as a progression through matter of a state of motion. The distinction between the progress of matter from one place to another, and the progress of a wave from one place to another through matter, is well illustrated by the very largest examples of waves that we have — largest in one dimension, smallest in another — waves of light, waves which extend from the remotest star, at least a million times as far from us as the sun is. Think of ninety-three million million miles, and think of waves of light coming from stars known to be at as great a distance as that! So much for the distance of propagation or progression of waves of light.

But there are two other magnitudes concerned in waves: there is the wavelength and there is the amount of displacement of a moving particle in the wave. Waves of light consist of vibrations to and fro, perpendicular to the line of progression of the wave. The length of the wave — I shall explain the meaning of "wavelength" presently: it speaks for itself in fact, if we look at waves of water — the length from crest to crest in waves of light is from one thirty-thousandth to one fifty-thousandth or one sixty-thousandth of an inch; and these waves of light travel through all known space. Waves of sound differ from waves of light in the

[133]Lecture delivered at the Conversazione of the Institution of Mechanical Engineers in the Science and Art Museum, Edinburgh, on Wednesday evening, 3rd August, 1887.

vibration of moving particles along the line of propagation of the wave, instead of perpendicular to it. Waves of water agree more nearly with waves of light than do waves of sound; but waves of water have this great distinction from waves of light and waves of sound, in that they are manifested at the surface or termination of the medium or substance whose motion constitutes the wave. It is with waves of water that we are concerned tonight; and of all the beautiful forms of water waves that of the Ship Waves is perhaps the most beautiful, if we can compare the beauty of such beautiful things. The subject of ship waves is certainly one of the most interesting in mathematical science. It possesses a special and intense interest, partly from the difficulty of the problem, and partly from the peculiar complexity of the circumstances concerned in the configuration of the waves.

Canal Waves. — I shall not at first speak of that beautiful configuration or wave pattern, which I am going to describe a little later, seen in the wake of a ship travelling through the open water at sea; but I shall as included in my special subject of ship waves, refer in the first place to waves in a canal, and to Scott Russell's splendid research on that subject, made about the year 1834 — fifty-three years ago — and communicated by him to the Royal Society of Edinburgh. I gave a very general and abstract definition of the term "wave"; let us now have it in the concrete: a wave of water produced by a boat dragged along a canal. In one of Scott Russell's pictures illustrating some of his celebrated experiments, is shown a boat in the position that he called behind the wave; and in the rear of the boat is seen a procession of waves. It is this procession of waves that we have to deal with in the first place. We must learn to understand the procession of waves in the rear of the canal boat, before we can follow, or take up the elements of the more complicated pattern which is seen in the wake of a ship travelling through open water at sea. Scott Russell made a fine discovery in the course of those experiments. He found that it is only when the speed of the boat is less than a certain limit that it leaves that procession of waves in its rear. Now the question that I am going to ask is: how is that procession kept in motion? Does it take power to drag the boat along, and to produce or to maintain that procession of waves? We all know it does take power to drag a boat through a canal; but we do not always think on what part of the phenomena manifested by the progress of the boat through the canal, the power required to drag the boat depends.

I shall ask you for a time to think of water not as it is, but as we can conceive a substance to be — that is, absolutely fluid. In reality, water is not perfectly fluid, because it resists change of shape; and non-resistance to change of shape is the definition of a perfect fluid. Is water then a fluid at all? It is a fluid because it permits change of shape; it is a fluid in the same sense that thick oil or treacle is a fluid. Is it only in the same sense? I say yes. Water is no more fluid in the abstract than is treacle or thick oil. Water, oil, and treacle all resist change of shape. When we attempt to make the change very rapidly, there is a great resistance; but if we make the change very slowly, there is a small resistance. The resistance of these fluids to change of shape is proportionate to the speed of the change: the quicker you change the shape, the greater is the force that is required to make the change. Only give it time, and treacle or oil will settle to its level in a glass or basin just as water does. No deviation from perfect fluidity, if the question of time does not enter, has ever been discovered in any of these fluids. In the case of all ordinary liquids, anything that looks like liquid and is transparent or clear — or, even if it is not transparent, anything that is commonly called a fluid or liquid — is perfectly liquid in the sense of exerting no permanent resistance to change of shape. The difference between water and a viscous substance, like treacle or oil, is defined merely by taking into account time. Now for some motions of water (as capillary waves), resistance to change of shape, or as we call it viscosity, has a very notable effect; for other cases viscosity has no sensible effect. I may tell you this — I cannot now prove it, for my function this evening is only to explain and bring before you generally some result of mathematical calculation and experimental observation on these subjects — I may tell you that great waves at sea will travel for hours or even for days, showing scarcely any loss of sensible motion – or of energy, if you will allow me so to call it— through viscosity. On the other hand, look at the ripples in a little pond, or in a little pool of fresh rain water lying in the street, which are excited by a puff of wind; the puff of wind is no sooner gone than the ripples begin to subside, and before you can count five or six the water is again perfectly still. The forces concerned in short waves such as ripples, and the forces concerned in long waves such as great ocean waves, are so related to time and to speed that, whereas in the case of short waves the viscosity which exists in water comes to be very potent, in the case of long waves it has but little effect.

Allow me then for a short time to treat water as if it were absolutely free from viscosity— as if it were a perfect fluid; and I shall afterwards endeavour to point out where viscosity comes into play, and causes the results of observation to differ more or less—very greatly in some cases, and very slightly in others — from what we should calculate on the supposition of water being a perfect fluid. If water were a perfect fluid, the velocity of progresion of a wave in a canal would be smaller the shorter the wave. That of a "long wave"—whose length from crest to crest is many times the depth of the canal—is equal to the velocity which a body acquires in falling from a height equal to half the depth of the canal. For brevity we might call this height the "speedheight" — the height from which a body must fall to acquire a certain speed. The velocity in feet per second is approximately eight times the square root of the height in feet. Thus in a canal 8 feet deep the natural velocity of the "long wave" is 16 feet per second, or about 11 miles per hour. If water were a perfect fluid, this would be the state of the case: a boat dragged along a canal at any velocity less than the natural speed of the long wave in the canal would leave a train of waves behind it of so much shorter length that their velocity of propagation would be equal to the velocity of the boat; and it is mathematically proved that the boat would take such a position as is shown in Scott Russell's diagram referred to, namely just on the rear slope of the wave. It was not by mathematicians that this was found out; but it was Scott Russell's accurate observation and well devised experiments that first gave us these beautiful conclusions.

To go back: a wave is the progression through matter of a state of motion. The motion cannot take place without the displacement of particles. Vary the definition by saying that a wave is the progression of displecement. Look at a field of corn on a windy day. You see that there is something travelling over it. That something is not the ears of corn carried from one side of the field to the other, but is the change of colour due to your seeing the sides or lower ends of the ears of corn instead of the tops. A laying down of the stalks is the thing that travels in the wave passing over the corn field. The thing that travels in the wave behind the boat is an elevation of the water at the crest and a depression in the hollow. You might make a wave thus. Place over the surface of the water in a canal a wave-form, made from a piece of pasteboard or of plastic material such as gutta-percha that you can mould to any given shape; and take care that the water fills up the wave-form everywhere, leaving no bubbles of air in the upper bends. Now you have a constant displacement of the water from its level. Now take your gutta-percha form and cause it to move along—drag it along the surface of the canal — and you will produce a wave. That is one of the best and most convenient of mathematical ways of viewing a wave. Imagine a wave generated in that way; calculate what kind of motion can be so generated, and you have not merely the surface motion produced by the force you applied, but you have the water motion in the interior. You have the whole essence of the thing discovered, if you can mathematically calculate from a given motion at the surface what is the motion that necessarily follows throughout the interior; and that can be done, and is a part of the elements of the mathematical results which I have to bring before you.

Now to find mathematically the velocity of progression of a free wave, proceed thus. Take your gutta-percha form and hold it stationary on the surface of the water; the water-pressure is less at the crest and greater at the hollow; by the law of hydrostatics, the deeper down you go, the greater is the pressure. Move your form along very rapidly, and a certain result, a centrifugal force, due to the inertia of the flowing water, will now cause the pressure to be greatest at the crest and least at the lowest point of the hollow. Move it along at exactly the proper speed and you will cause the pressure to be equal all over the surface of the gutta-percha form. Now have done with the gutta-percha. We only had in it imagination. Having imagined it and got what we wanted out of it, discard it. When moving it at exactly this proper speed, you have a free wave. That is a slight sketch of the mode by which we investigate mathematically the velocity of the free wave. It was by observation that Scott Russell found it out; and then there was a mathematical verification, not of the perfect theoretic kind, but of a kind which showed a wonderful grasp of mind and power of reasoning upon the phenomena that he had observed.

But still the question occurs to everybody who thinks of these things in an engineering way, how does that procession require work to be done to keep it up? or does it require work to be done at all? May it not be that the work required to drag the boat along the canal has nothing to do with the waves after all? that the formation of the procession of waves once effected leaves nothing more to be desired in the way of work? that the procession once formed will go on of itself, requiring no work to sustain it? Here is the explanation. The procession has an end. The canal may be infinitely long, the time the boat may be going may be as long as you please; but let us think of a beginning– the boat started, the procession began to form. The next time you make a passage in a steamer, especially in smooth water, look behind the steamer, and you will see a wave or two as the steamer gets into motion. As it goes faster and faster, you will see a wave-pattern spread out; and if you were on shore, or in a boat in the wake of the steamer, you would see that the rear end of the procession of waves follows the steamer at an increasing distance behind. It is an exceedingly complicated phenomenon, and it would take a great deal of study to make out the law of it merely from observation. In a canal the thing is more simple. Scott Russell however did not include this in his work. This was left to Stokes, to Osborne Reynolds, and to Lord Rayleigh. The velocity of progress of a wave is one thing; the velocity of the front of a procession of waves, and of the rear of a procession of waves, is another thing. Stokes made a grand new opening, showing us a vista previously unthought of in dynamical science. As was his manner, he did it merely in an examination question set for the candidates for the Smith prize in the University of Cambridge. I do not remember the year, and I do not know whether any particular candidate answered the question; but this I know, that about two years after the question was put, Osborne Reynolds answered it with very good effect indeed. In a contribution to the Plymouth Meeting of the British Association in 1877 (see Nature 23 Aug. 1877, pages 343-4), in which he worked out one great branch at all events of the theory thus pointed out by Stokes, Reynolds gave this doctrine of energy that I am going to try to explain; and a few years later Lord Rayleigh took it up and generalised it in the most admirable manner, laying the foundation not only of one part, but of the whole, of the theory of the velocity of groups of waves.

The theory of the velocity of groups of waves, on which is founded the explanation of the wave-making resistance to ships whether in a canal or at sea, I think I have explained in such a way that I hope every one will understand the doctrine in respect to waves in a canal; it is more complex in respect to waves at sea. I shall try to give you something on that part of the subject; but as to the dynamical theory, you will see it clearly in regard to waves in a canal. If Scott Russell's drawing were continued backwards far enough, it would show an end to the procession of waves in the rear of the boat; and the distance of that end would depend on the time the boat had been travelling. You will remember that we have hitherto been supposing water to be free from viscosity; but in reality water has enough of viscosity to cause the cessation of the wave procession at a distance corresponding to 50 or 60 or 100 or 1000 wavelengths in the rear of the ship. In a canal especially viscosity is very effective, because the water has to flow more or less across the bottom and up and down by the banks; so that we have not there nearly the same freedom that we have at sea from the effects of viscosity in respect to waves. The rear of the procession travels forward at half the speed of the ship, if the water be very deep. What do I mean by very deep? I mean a depth equal to at least one wave-length; but it will be nearly the same for the waves if the depth be three-quarters of a wave-length. For my present purpose in which I am not giving results with minute accuracy, we will call very deep any depth more than three-quarters of the wave-length. For instance, if the depth of the water in the canal is anything more than three-quarters of the length from crest to crest of the waves, the rate of progression of the rear of the procession will be half the speed of the boat. Here then is the state of the case. The boat is followed by an ever-lengthening procession of waves; and the work required to drag the boat along in the canal— supposing that the water is free from viscosity — is just equal to the work required to generate the procession of waves lengthening backwards behind the boat at half the speed of the boat. The rear of the procession travels forwards at half the speed of the boat; the procession lengthens backwards relatively to the boat at half the speed of the boat. There is the whole thing; and if you only know how to calculate the energy of a procession of waves, assuming the water free from viscosity, you can calculate the work which must be done to keep a canal boat in motion.

But now note this wonderful result: if the motion of the canal boat be more rapid than the most rapid possible wave in the canal (that is, the long wave), it cannot leave behind it a procession of waves—it cannot make waves, properly so called, at all; it can only make a hump or a hillock travelling with the boat, as shown in another of Scott Russell's drawings. What would you say of the work required to move the boat in that case? You may answer that question at once:it would require no work; start it, and it will go on for ever. Every one understands that a curling stone projected along the ice would go on for ever, were it not for the friction of the ice; and therefore it must not seem so wonderful that a boat started moving through water would also go on for ever, if the water were perfectly fluid: it would not, if it is forming an ever-lengthening procession of waves behind it; it would go on for ever, if it is not forming a procession of waves behind it. The answer then simply is, give the boat a velocity greater than the velocity of propagation of the most rapid wave (the long wave) that the canal can have; and in these circumstances, ideal so far as nullity of viscosity is concerned, it will travel along and continue moving without any work being done upon it. I have said that the velocity of the long wave in a canal is equal to the velocity which a body acquires in falling from a height equal to half the depth of the canal. The term "long wave" I may now further explain as meaning a wave whose length is many times the depth of the water in the canal—50 times the depth will fulfil this condition—the length being always reckoned from crest to crest. Now if the wave-length from crest to crest be 50 or more times the depth of the canal, then the velocity of the wave is that acquired by a body falling through a height equal to half the depth of the canal; if the wavelength be less than that, the velocity can be expressed only by a complex mathematical formula. The results have been calculated; but I need not put them before you, because we are not going to occupy ourselves with them.

The conclusion then at which we have arrived is this: supposing at first the velocity of the boat to be such as to make the waves behind it of wave-length short in comparison with the depth of water in the canal: let the boat go a little faster, and give it time until steady waves are formed behind it; these waves will be of longer wavelength: the greater the speed of the boat, the longer will be the wavelength, until we reach a certain limit; and as the wavelength begins to be equal to the depth, or twice the depth, or three times the depth, we approach a wonderful and critical condition of affairs—we approach the case of constant wave velocity. There will still be a procession of waves behind the boat, but it will be a shorter procession and of higher waves; and this procession will not now lengthen astern at half the speed of the boat, but will lengthen perhaps at a third, or a fourth, or perhaps at a tenth of the speed of the boat. We are approaching the critical condition: the rear of the procession of waves is going forward nearly as fast as the boat. This looks as if we were coming to a diminished resistance; but it is not really so. Though the procession is lengthening less rapidly relatively to the boat than when the speed was smaller, the waves are very much higher; and we approach almost in a tumultuous manner to a certain critical velocity. I will read you presently Scott Russell's words on the subject. Once that crisis has been reached, away the boat goes merrily, leaving no wave behind it, and experiencing no resistance whatever if the water be free from viscosity, but in reality experiencing a very large resistance, because now the viscosity of the water begins to tell largely on the phenomena. I think you will be interested in hearing Scott Russell's own statement of his discovery. I say his discovery, but in reality the discovery was made by a horse, as you will learn. I found almost surprisingly in a mathematical investigation, "On Stationary Waves in Flowing Water", contributed to the Philosophical Magazine (Oct., Nov., Dec., 1886, and Jan., 1887), a theoretical confirmation, forty-nine and a half years after date, of Scott Russell's brilliant "Experimental Researches into the Laws of Certain Hydrodynamical Phenomena that accompany the Motion of Floating Bodies, and have not previously been reduced into conformity with the known Laws of the Resistance of Fluids."[134]

These experimental researches led to the Scottish system of fly-boats carrying passengers on the Glasgow and Ardrossan Canal, and between Edinburgh and Glasgow on the Forth and Clyde Canal, at speeds of from eight to thirteen miles an hour, each boat drawn by a horse or pair of horses galloping along the bank. The method originated from the accident of a spirited horse, whose duty it was to drag the boat along at a slow walking speed, taking fright and running off, drawing the boat after him; and it was discovered that, when the speed exceeded the velocity acquired by a body falling through a height equal to half the depth of the canal (and the horse certainly found this), the resistance was less than at lower speeds. Scott Russell's description of how Mr. Houston took advantage for his Company of the horse's discovery is so interesting that I quote it in extenso:

> Canal navigation furnishes at once the most interesting illustrations of the interference of the wave, and most important opportunities for the application of its principles to an improved system of practice. It is to the diminished anterior section of displacement, produced by raising a vessel with a sudden impulse to the summit of the progressive wave, that a very great improvement recently introduced into canal transports owes its existence. As far as I am able to learn, the isolated fact was discovered accidentally

---

[134] By John Scott Russell, Esq., M.A., F.R.S.E. Read before the Royal Society of Edinburgh, 3rd April, 1837, and published in the Transactions in 1840.

on the Glasgow and Ardrossan Canal of small dimensions. A spirited horse in the boat of William Houston, Esq., one of the proprietors of the works, took fright and ran off dragging the boat with it, and it was then observed, to Mr. Houston's astonishment, that the foaming stern surge which used to devastate the banks had ceased, and the vessel was carried on through water comparatively smooth with a resistance very greatly diminished. Mr. Houston had the tact to perceive the mercantile value of this fact to the canal company with which he was connected, and devoted himself to introducing on that canal vessels moving with this high velocity. The result of this improvement was so valuable, in a mercantile point of view, as to bring, from the conveyance of passengers at a high velocity, a large increase of revenue to the canal proprietors. The passengers and luggage are conveyed[135] in light boats, about sixty feet long and six feet wide, made of thin sheet iron, and drawn by a pair of horses. The boat starts at a slow velocity behind the wave, and at a given signal it is by a sudden jerk of the horses drawn up on the top of the wave, where it moves with diminished resistance, at the rate of 7, 8 or 9 miles an hour."

Scott Russell was not satisfied with a mere observation of this kind. He made a magnificent experimental investigation into the circumstances. An experimental station at the Bridge of Hermiston on the Forth and Clyde Canal was arranged for the work. It was so situated that there was a straight run of 1500 feet along the bank, and, in the drawing of it in Scott Russell's paper, three pairs of horses are seen galloping along. They seem to be galloping on air, but are of course on the towing path; and this remark may be taken as an illustration that, if the horses only galloped fast enough, they could gallop over the water without sinking into it, as they might gallop over a soft clay field. That is a sober fact with regard to the theory of waves; it is only a question of time how far the heavy body will enter into the water, if it is dragged very rapidly over it. This, however, is a digression. In the very ingenious apparatus of Scott Russell's, there is a pyramid 75 feet high, supporting a system of pulleys which carry a heavy weight suspended by means of a rope. The horses are dragging one end of this rope, while the other end is fastened to a boat which travels in the opposite direction. It is the old principle invented by Huyghens, and still largely used, in clockwork. Scott Russell employed it to give a constant dragging force to the boat from the necessarily inconstant action of the horses. I need not go into details, but I wish you to see that Scott Russell, in devising these experiments, adopted methods for accurate measurement in order to work out the theory of those results, the general natural history of which he had previously observed.

---

[135]This statement was made to the Royal Society of Edinburgh in 1837, and it appeared in the Transactions in 1840. Almost before the publication in the Transactions the present tense might, alas! have been changed to the past —"passengers were conveyed." Is it possible not to regret the old fly-boats between Glasgow and Ardrossan and between Glasgow and Edinburgh, and their beautiful hydrodynamics, when, hurried along on the railway, we catch a glimpse of the Forth and Clyde Canal still used for slow goods traffic; or of some swampy hollows, all that remains of the Ardrossan Canal on which the horse and Mr. Houston and Scott Russell made their discovery?

Appendix II
Skyrme's Soliton

Tony Hilton Royle Skyrme (1922–1987) was born on 5 December 1922 in London. His parents were John Hilton Royle Skyrme, a bank clerk, and Muriel May née Roberts. Tony's maternal grandfather was Herbert William Thomson Roberts, by profession a calculator of tides for the Admiralty. The inclusion of Lord Kelvin's baptismal name (William Thomson) among his forenames reflects the professional contact that Tony's great grandfather had with Lord Kelvin and the high regard in which he held the latter.

Tony's great grandfather on his maternal side was Edward Roberts, born in 1845. In 1868, he was appointed Secretary to the Tidal Committee for the Advancement of Science, and was made responsible later for construction of the first Tidal Predicter, which had been designed by Lord Kelvin for this Committee, and which was constructed during the winter months of 1872/73 and first exhibited at the Bradford Meeting of the British Association in 1873. Later, he played a large part in the design and construction of the Universal Tide-predicting Machines used by the Indian and Colonial Governments and by the Admiralty Hydrographic Office.

It was his house that held the Tidal Predicter, the first model of the machine, which made such a strong impression on the young Tony and influenced so greatly his later ideas, as Tony himself recounted in a lecture given at a Workshop on Skyrmions in 1984.

Tony was a theoretical physicist of unusual imagination, but with a deep sensitivity and abnormally strong degree of self-criticism. His mathematical perceptions were very deep, and he could see layers deeper than anyone else in his locale. He was unusually quick in understanding, on any topic. Professionally, although mathematical in style, he was in close touch with physical significance in his work.

R.H. Dalitz

[135] Adapted from the paper "An outline of the life and work of Tony Hilton Royle Skyrme (1922–1987)," *Int. Jour. of Modern Physics A3*(1988) 2719.

T.H.R. Skyrme
THE ORIGINS OF SKYRMIONS

This talk was given as a historical introduction to the Workshop on ""Skyrmions" held at Cosener's House, Abingdon, 17–18 November 1984. It has been re-constructed by Dr. Ian Aitchison from sketchy notes left by Tony Skyrme, and from his own notes taken at that talk, aided by notes taken by several others who were present.

Published in the same journal (the same issue) as the biographical note quoted above. It was also published in *A Breadth of Physics*, eds. R.H. Dalitz, R.B. Stinchcombe (World Scientific, Singapore, 1988).

I had three main motives in trying to make a model of the kind we are discussing: unification, the renormalisation problem, and what I call the "fermion problem." The first of these is fairly obvious. Unification of one kind or another has always been a goal of theoretical physicists. I know that some of our colleagues are suspicious of attempts to bring in grand ideas of symmetry and unification for their own sake, unless they are clearly required by the experimental evidence. However, I believe that most progress in physics has been due to ideas of this kind — though it is equally true that many of the ideas have been wild speculation. In the present context, this philosophy led me to think that instead of having two fundamental types of particle — bosons and fermions — it would be nice if we only had one. Heisenberg had proposed his non-linear spinor theory in 1958, in which everything was made from one fundamental self-interacting fermion field. For reasons which I will explain in a moment, I didn't like fermions much, and thought it would be fun to see if I could get everything out of a self-interacting boson field theory instead!

The other two "problems" are of course very modern ones, and specific to quantum field theory; but in thinking about them, I have noticed that they have appeared earlier in the history of physical ideas, in rather different guises. In particular, I have found it interesting to go back to the views of that eminent Victorian, Sir William Thomson (later Lord Kelvin (1824–1907)). Kelvin was deeply concerned with the problems of the atomic structure of matter, and its bearing on the theory of gases, and so on, just as today we are concerned with the elementary particle structure of nuclear matter.

Kelvin was very reluctant to accept the idea of infinitely rigid point-like atoms. In one of his lectures he spoke of

> ... the monstrous assumption of infinitely strong and infinitely rigid pieces of matter, the existence of which is asserted as a probable hypothesis by some of the greatest modern chemists in their rashly-worded introductory statements. ...

He seems to have felt intuitively that there was something deeply wrong with the idea—though I am not sure precisely why. Anyway, I have always found the idea of any sort of elementary particle as a point-like object unreasonable — and of course we have good reason for uneasiness, because such a theory has no natural cut off and infinite renormalisation seems inevitable. Certainly, renormalisation theory has been built, through the efforts of a number of distinguished physicists, into a beautiful and ingenious form; but I still feel that it is just a very good and useful way of enabling us to live with our ignorance of what really goes on at short distances. Such accommodation should only be provisional. Indeed, these problems would not arise in the first place if the fundamental particles were actually extended objects.

The infinite divisibility of matter seemed absurd to the Greek philosophers of the Epicurean School, and to Lucretius who gives such a fine account of their ideas. This was one of their reasons for believing in atoms. Another was their feeling that there had to be some kind of atomic structure to conserve specific qualities. As Kelvin put it in the same lecture already quoted:

> For the only pretext seeming to justify the monstrous assumption of ...
> their rashly-worded introductory statements, is that urged by Lucretius and
> adopted by Newton; that it seems necessary to account for the unalterable
> distinguishing qualities of different kinds of matter.

This was the point seized on by Kelvin, and given a quite different explanation in terms of "vortex atoms," inspired by Helmholtz's work in fluid motion. Kelvin continued his lecture thus:

> But Helmholtz has proved an absolutely unalterable quality in the motion
> of any portion of a perfect liquid, in which the peculiar motion which he calls
> *Wirbelbewegung* has been once created. Thus, any portion of a perfect liq-
> uid which has *Wirbelbewegung* has one recommendation of Lucretius' atoms
> — infinitely perennial specific quality. To generate or destroy "Wirbelbewe-
> gung" in a perfect fluid can only be an act of creative power. Lucretius' atom
> does not explain any of the properties of matter without attributing them to
> the atom itself. Thus, the "clash of atoms" as it has been well called, has been
> invoked by his modern followers to account for the elasticity of gases. Ev-
> ery other property of matter has similarly required an assumption of specific
> forces pertaining to the atom. It is as easy (and as improbable, if not more
> so) to assume whatever specific forces may be required in any portion of
> matter which possesses the *Wirbelbewegung* as in a solid indivisible piece of
> matter, and hence the Lucretius atom has no prima facie advantage over the
> Helmholtz atom. A magnificent display of smoke-rings, which he recently
> had the pleasure of witnessing in Professor Tait's lecture room, diminished
> by one the number of assumptions required to explain the properties of mat-
> ter, on the hypothesis that all bodies are composed of vortex-atoms in a per-
> fect homogeneous liquied. Two smoke-rings were frequently seen to bound
> obliquely from one another, shaking violently from the effects of the shock.

I shall come back to this "vortex" idea in a moment, but before that I turn to the third problem—the "fermion" one. I know that the idea of an intrinsic fermion is perfectly acceptable to many—perhaps most—people, but I have always felt an unease about quantum-mechanical concepts that do not have clear classical analogues. Now that quantum mechanics can be well understood as an averaging over classical configurations, it seems even more slightly anomalous. Fundamental fermions are awkward to handle in the path integral formalism. Admittedly, we can incorporate them by using Grassmann variables, but this seems an unnatural, purely mathematical, construction. I would like to think that the fermion concept was just a good way of talking about the behavior of some semi-classical construction, and that it was no more fundamental than renormalisation.

Here we come back to Kelvin. Almost 100 years ago, he gave a series of lectures at Johns Hopkins University — for which he was offered expenses and $1000, quite a substantial sum for those days. He subsequently wrote the lectures up, and published them as the Baltimore Lectures. Kelvin was reluctant, for much of his life, to accept unreservedly the ideas of Maxwell, his contemporary. In one of his Baltimore lectures, he tried to explain why:

> I can never satisfy myself until I can make a mechanical model of a thing. If I can make a mechanical model, I can understand it. As long as I cannot make a mechanical model all the way through I cannot understand; and that is why I cannot get [this is probably the reporter's Americanism for the word "accept"] the electromagnetic theory. I firmly believe in an electromagnetic theory of light, and that when we understand electricity and magnetism and light we shall see them all together as parts of the whole. But I want to understand light as well as I can, without introducing things that we understand even less of. That is why I take plain dynamics. I can get a model in plain dynamics; I cannot in electromagnetics. But as soon as we have rotators to take the part of magnets, and something imponderable to take the part of magnetism, and realise by experiment Maxwell's beautiful ideas of electric displacements and so on, then we shall see electricity, magnetism, and light closely united and grounded in the same system.

For Kelvin, then, an "understanding" meant a mechanical model. He and Tait spent a long time developing "smoke ring" or "vortex" models of the atom (based on Helmholtz's work), and also various mechanical models of the luminiferous ether.

Later in life Kelvin came to accept Maxwell's theory; and his own grand design of vortex atoms had to be abandoned. Sadly, he felt that his life had been a failure, despite his many brilliant achievements.

Analogies with the subject we are discussing today are fairly evident, though I hesitate to draw any moral from history. Certainly, I can remember being greatly impressed by a machine he and Tait had caused to be built in 1873 for the Tidal Committee of the British Association, for predicting the tides world-wide. This machine was in my grandfather's house, and the ingenuity of its mechanism, whereby it could produce this complicated pattern of tides, had considerable influence upon me. Anyway, I wanted a physical model which would reproduce the curious behavior of fermions. I had been vaguely aware of Kelvin's picture

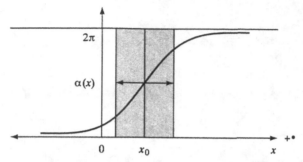

A.1 A simple kink centered on $x = x_0$.
The arrows show its width, the x-range over which
most of the variation of $\alpha(x)$ occurs.

of vortex atoms, and I had always been attracted by the type of structure that was possible in non-linear theories such as general relativity, or the Born–Infeld theory of electromagnetism. I liked the idea, for example, that the "sources" of gravitation might themselves be produced by the field equations—as, presumably, some kind of singularities in the fields—instead of having to be put in by hand. There was here an obvious hope that somehow fermionic sources of "strong charge" or baryon number might emerge as singularities of some non-linear classical meson field theory. . . .

To simplify the problem I looked at the analogous problem in one space (and one time) dimension—the problem now known as the Sine-Gordon equation. This employs a single angle-type field variable $\alpha(x, t)$, in one space dimension, and the equation of motion is

$$\partial_t^2 - \partial_x^2 = -m^2 \sin \alpha.$$

A vacuum state is such that $\cos \alpha = 1$, but of course we can have a situation in which a time independent $\alpha(x)$ interpolates from the vacuum at $x = -\infty$ (e.g., $\alpha(-\infty) = 0$) to another at $x = +\infty$ (e.g. $\alpha(+\infty) = 2\pi$), as shown in the figure below.

The "kink" or singularity propagates according to the one-dimensional "neutrino-like" equation

$$(\partial_t - \partial_x)\psi(x, t) = 0,$$

and ψ obeys anticommutation relations. . . .

I still entertain the hope that some type of non-linear theory will yield an explanation of elementary particles that can be visualised in a semiclassical way, and that the quarks or leptons introduced as sources in most theories will be seen to be mathematical constructs helpful in its understanding, rather than fundamental constituents, just as the idea of vortices in a fluid is an indispensible way of talking about certain types of fluid motion.

Mathematics Appendix

Here we first present some nonlinear equations and their simplest solitonic solutions and then give a brief summary of the main events in mathematical theory of solitons.

1. The KdV equation was written in Chapter 7 (eqs.(7.1) and (7.2)). Usually one writes it for the dimensionless function $u \equiv 3y/4h$ depending on the dimensionless time and space variables $T \equiv \sqrt{6}v_0 t/h$ and $X \equiv \sqrt{6}x/h$:

$$\dot{u} + (u + u^2 + u'')' = 0.$$

Here the dot denotes the derivative with respect to "time" T and the prime denotes the "space" derivative (with respect to X). The one-soliton solution of this equation is

$$u = 6k/\cosh^2[k(X - VT)],$$

where k is an arbitrary real number and $V = 1 + 4k^2$ (compare this to eqs. (7.1) and (7.2)). By replacing in the KdV equation u^2 with u^3, we get the so-called "modified" KdV equation. Its one-soliton solution is

$$u = \sqrt{2}k/\cosh[k(X - VT)], \quad V = 1 + k^2.$$

The **sine-Gordon equation** was written in Chapter 6 (eq. (6.11)). One usually writes it for the function $u = \pi + \phi$ of the dimensionless variables $T = \omega_0 t$ and $X = \omega_0 x/v_0$:

$$u'' - \ddot{u} = \sin u.$$

Its one-soliton solution was also written in Chapter 6:

$$u = 4 \arctan \exp[\beta(X - VT)], \quad \beta = 1/\sqrt{1 - V^2}.$$

Two solitons are described by the solution

$$u = 4 \arctan[V \sinh(\beta x)/\tanh(\beta V T)].$$

The soliton-antisoliton solution is

$$u = 4 \arctan[V^{-1} \sinh(\beta V T)/\cosh(\beta x)],$$

and the breather solution is

$$u = 4 \arctan[a \sin(bT)/b \cosh(aX)], \quad a^2 + b^2 = 1.$$

Typical solitons in discrete lattices are the solitons in the nonlinear **Toda lattice**. The system of equations describing movements of "atoms" in the Toda lattice is the following:

$$\ddot{u}_n = \exp(u_{n+1} - u_n) - \exp(u_n - u_{n-1}),$$

where u_n are the discrete (dimensionless) coordinates of the atoms. The solitonic solution of the Toda system is given by the formulae

$$u_n = s_n - s_{n+1}, \quad s_n = \ln\{1 + \exp[2(an + T \sinh a)]\},$$

where a is an arbitrary real number. Note that the discretized KdV equation has the form

$$\dot{u}_n = \exp u_{n+1} - \exp u_{n-1},$$

while the continuum limit of the Toda equations is the Boussinesq equation

$$\ddot{u} = (u + u^2 + u'')'',$$

which sometimes is called the nonlinear string equation.

Finally, let us write a typical equation describing **nonlinear diffusion**

$$u'' - \dot{u} = u(u - 1)(u - a)$$

and its solitary wave solution

$$u = \left[1 + \exp[(X - VT)/\sqrt{2}]\right]^{-1}, \quad V = (1 - 2a)/\sqrt{2}.$$

Other interesting equations describing different solitons may be found in the book by M. Ablowitz and H. Segur: *Solitons and the Inverse Scattering Transform*, SIAM, Philadelphia, 1981. The main topic of this book is the inverse scattering method in the theory of solitons but the authors also describe many physical applications of solitons.

2. A systematic development of the mathematical theory of solitons began in 1967 when C. Gardner, J. Green, M. Kruskal and R. Miura proposed the method of solving the KdV equation based on the so-called "inverse scattering theory." The

following year, P. Lax essentially generalized this method. In 1971, V.E. Zakharov and A.B. Shabat applied the inverse scattering method (in the form given by Lax) to different types of solitonic equations, including the nonlinear Schrödinger equation. In the same year, V.E. Zakharov and L.D. Faddeev proved that the KdV equation, which they treated as an infinite-dimensional Hamiltonian system, is completely integrable. All the mentioned papers used a rather complex mathematical apparatus of the inverse scattering method, which allows one to reduce integrable nonlinear partial differential equations to linear ones. In 1971, a simpler direct approach to constructing solitonic solutions of different integrable equations was proposed by R. Hirota.

In 1973, M. Ablowitz, D. Kaup, A. Newell and H. Segur started a systematic study of integrable equations and of their solitonic solutions. In particular, the complete integrability of the sine-Gordon equation was proved. In 1974–1975, S.P. Novikov and his school found a general approach to constructing exact periodic solutions of the KdV equation, using deep mathematical results of Riemann, Abel and Jacobi as well as their modern generalizations. A further development of these ideas resulted in discovering deep relations between superstrings and solitons (and with integrable systems in general). This happened later and was achieved by joint efforts of many mathematicians and theoretical physicists working as one team.

Today, most interesting ideas in soliton theory are connected with quantum integrability and quantum theory of solitons. This field of science is being explored equally by mathematicians and theoretical physicists. Quantum solitons have numerous applications, especially, in condensed matter physics. However, there are more and more hints that soliton ideas are also very useful in studies of the fundamental structure of the world. For example, the story of the magnetic monopole and of the electric-magnetic duality has recently been continued. American theorists Nathan Seiberg and Edward Witten vigorously started investigations into new dualities both in gauge and string theories (see e.g., E. Witten's popular review in *Physics Today*, May 1997). Many mathematically oriented theorists are now engaged in studies of dualities in different models, hoping that they will result in new insights into the structure of the real world. The ideas and methods developed in the mathematical theory of solitons play a significant role in theses studies.

The "classical" period of the mathematical theory of solitons is summarized in the book: A.C. Newell, *Solitons in Mathematics and Physics*, SIAM, Philadelphia, 1985. The quantum story is not yet described in books and the interested reader should consult current scientific literature or databases (e.g. http://xyz.lanl.gov).

Another reference on KdV is "KdV-Centennial for the publication of the equation named after Korteweg and de Vries." Proceedings of Symposium, Kluwer, 1995, ed. M. Hazenwinkel.

I also recommend two review papers on applications of the soliton ideas: 1) N. J. Zabusky, "Computational Synergetics and Mathematical Innovation," *Journ. of Comp. Phys.*, Vol. 43, N2, Oct. 1981, pp. 195–249; 2) A. C. Scott, "The Electrophysics of a Nerve Fiber," *Rev. Mod. Phys.*, Vol. 47, N2, April 1975, pp. 487–533.

Glossary of Names

The following represents the names of people quoted in the book.

Abel, Niels Henrik (1802–1829)
Abrikosov, Aleksei Alekseevich (1928)
Adams, John Couch (1819–1892)
Airy, George Biddell (1801–1892)
Allen, John Frank (1908)
Ampère, André Marie (1775–1836)
Anderson, Philip Warren (1923)
Andronov, Aleksandr Aleksandrovich (1901–1952)
Arago, Dominique François Jean (1786–1853)
Aristotle (384–322 BC)
Arnold, Vladimir Igorevich (1937)
Atiyah, [Sir] Michael Francis (1929)
Avogadro, Amedeo (1776–1856)
Babbage, Charles (1792–1871)
Bardeen, John (1908–1991)
Beethoven, Ludwig van (1770–1827), German composer
Belinskii, Vissarion Grigorievich (1810–1842), Russian critic, essayist
Benjamin, T. Brook
Bergmann, Peter Gabriel
Bernoulli, Daniel (1667–1748)
Bernoulli, Johann (1700–1782)
Bernoulli, Jacob (1654–1705)
Bernstein, Julius (1839–1917)
Biot, Jean Baptiste (1774–1862)
Bitter, Francis (1902–1967)

Blake, William (1751–1827), English poet
Bloch, Felix (1905–1983)
Blok, Aleksandr Aleksandrovich (1880–1921), Russian poet
Bogoliubov, Nikolai Nikolaevich (1909–1993)
Bohr, Niels Henrik David (1885–1962)
Boltzmann, Ludwig (1844–1906)
Bolyai, Janos (1802–1860)
Bolzano, Bernhard (1781–1848)
Bonaparte, Napoleon (Napoleon I, Emperor of France) (1769–1821)
Boole, George (1815–1864)
Born, Max (1882–1970)
Bose, Satyendra Nath (1894–1974)
Boussinesq, Joseph Valentine de (1842–1929)
Boyle, Robert (1627–1691)
Bragg, [Sir] William Henri (1862–1942)
Bragg, [Sir] William Lawrence (1890–1971)
Broglie, Louis Victor de (1892–1987)
Bryussov, Valerii Yakovlevich (1873–1924), Russian poet
Bullough, Robert Keith (1929)
Byron, George Gordon Noel (1788–1824)
Cabibbo, Nicola (1935)
Callan, Curtis (1942)
Carlyle, Thomas (1795–1881), English historian and essayist
Carnot, Nicolas Leonard Sadi (1796–1832)
Cauchy, Augustin Louis (1789–1857)
Cavendish, Henri (1731–1810)
Cayley, Arthur (1821–1895)
Champollion, Jean-François (1790–1832), French Egyptologist
Chladni, Ernst Friedrich (1756–1827)
Clapeyron, Benoit Pierre Emile (1799–1864)
Clausius, Rudolf Julius Emanuel (1822–1888)
Cole, Kenneth Stewart (1900)
Colladone, Jean Daniel (1802–1892)
Compton, Arthur Holly (1892–1962)
Comte, Auguste (1798–1857), French philosopher
Cook, William (1806–1879)
Cooper, Leon (1930)
Coulomb, Charles Augustin de (1736–1806)
Crelle, August Leopold (1780–1855),
D'Alembert, Jean Le Rond (1717–1783)
Darwin, Charles Robert (1808–1882)
Davisson, Clinton Joseph (1881–1958)
Deem, Gary S.
Descartes, René du Perron (1596–1650)
de Vries, Gustav
Diderot, Denis (1713–1784), French philosopher and writer
Dietrich, I
Dirac, Paul Adrien Maurice (1902–1984)
Dorfman, Yakov Grigorievich (1898–1974)

Du Bois-Reymond, Emil (1818–1896)
Eco, Umberto (1932), Italian writer
Ehrenfest, Paul (1880–1933)
Einstein, Albert (1879–1955)
Emmerson, George S.
Esaki, Leo (1925)
Euclid [of Alexandria] (ca 340 or ca 365–ca 287 or ca 300 BC)
Euler, Leonhard (1707–1783)
Ewing, James Alfred (1855–1935)
Faddeev, Ludwig Dmitrievich (1934)
Faraday, Michael (1791–1867)
Feigenbaum, Mitchell
Feir, J.E.
Fermi, Enrico (1901–1954)
Fet, Aphanasii Aphanasievich (1820–1892), Russian poet
Feynman, Richard Phillips (1918–1988)
Fock, Vladimir Aleksandrovich (1898–1974)
Forbes, James David (1804–1868)
Foucault, Jean Bernard Leon (1819–1868)
Fourier, Jean Baptiste Joseph (1768–1830)
Frank-Kamenetskii, David Albertovich (1910–1970)
Frenkel, Victor Yakovlevich (1929)
Frenkel, Yakov Ilyich (1894–1952)
Fresnel, Augustin Jean (1788–1827)
Friedrichs, Kurt Otto (1901)
Galilei, Galileo (1564–1642)
Galoise, Evariste (1811–1832)
Galvani, Luigi (1737–1798)
Gardner, Clifford S.
Gauss, Karl Friedrich (1777–1855)
Gell-Mann, Murray (1929)
Germer, Lester Halber (1896–1971)
Gerstner, Frantischek Josef (1756–1832)
Giaver, Ivar (1929)
Gibbs, Josiah Willard (1839–1903)
Ginzburg, Vitalii Lazarevich (1916)
Glashow, Sheldon (1932)
Goethe, Johann Wolfgang von (1749–1832)
Gordon, Walter (1893–1939)
Gordon, J.P.
Gorky, Maksim (Peshkov, Aleksei Maksimovich) (1868–1936), Russian writer
Grassmann, Hermann Gunther (1809–1877)
Green, George (1793–1841)
Greene, John M. (1928)
Green, Michael B.
Hamilton, William Rowan (1805–1865)
Hasegawa, Akira (1934)
Heine, Heinrich (1797–1856), German poet
Heisenberg, Werner Karl (1901–1976)

Helmholtz, Hermann (1821–1894)
Henri, Joseph (1797–1878)
Heraclitus [of Ephesus] (ca 550–475 BC)
Hermann, Ludimar (1838–1914)
Hershel, John Frederik William (1792–1871)
Hertz, Heinrich Rudolf (1857–1894)
Higgs, Peter Ware (1929)
Hilbert, David (1862–1943)
Hirota, R.
Hodgkin, Alan Lloyd (1914)
Holm, R.
Hooke, Robert (1635–1703)
Huxley, Aldous (1894–1963), English writer
Huxley, Andrew Fielding (1917)
Huyghens, Christiaan (1629–1695)
Hyers, Donald H.
Infeld, Leopold (1898–1968)
Ioffe, Abram Fyodorovich (1880–1960) (also, Joffe)
Iwanenko, Dmitrii Dmitrievich (1904–1994)
Jacobi, Karl Gustav Jacob (1804–1851)
Jordan, Ernst Pascual (1902–1980)
Josephson, Brian (1940)
Joule, James Prescott (1818–1889)
Kadomtsev, Boris Borisovich (1928)
Kaluza, Theodore Franz Eduard (1885–1954)
Kammerlingh Onnes, see Onnes
Kapitsa, Pyotr Leonidovich (1894–1984)
Kelvin, [Lord] William (1824–1907) (William Thomson)
Kepler, Johannes (1571–1630)
Kerr, John (1824–1907)
Khaikin, Semyon Emmanuilovich (1901–1968)
Kirchhoff, Gustav Robert (1824–1887)
Klein, Oskar Benjamin (1894–1977)
Kolmogorov, Andrei Nikolaevich (1903–1987)
Kompaneets, Aleksandr Solomonovich (1914–1974)
Kontorova, T.A.
Korteweg, Diderik Johannes (1848–1941)
Kruskal, Martin David (1925)
Krutkov, Yurii Aleksandrovich (1890–1952)
Krylov, Aleksei Nikolaevich (1883–1945)
Krylov, Nikolai Mitrofanovich (1879–1955)
Lagrange, Joseph Louis (1736–1813)
Lamb, Horace (1849–1934)
Landau, Lev Davidovich (1908–1968)
Laplace, Pierre Simon, Marquis de (1749–1827)
Larmor, Joseph (1847–1942)
Lavrentiev, Mikhail Alekseevich (1900–1980)
Lax, Peter David (1926)
Lazarev, Pyotr Petrovich (1878–1942)

Lebedev, Pyotr Nikolaevich (1866–1912)
Leibniz, Gottfried Wilhelm (1646–1717)
Lenz, Emilii Christianovich (1804–1865)
Leoté, Henri Charles Victor (1847–1916)
Le Verrier, *see* Verrier
Lewis, Gilbert Newton (1875–1946)
Lifshits, Evgenii Mikhailovich (1915–1985)
Lobachevskii, Nikolai Ivanovich (1792–1856)
Lomonosov, Mikhail Vassilievich (1711–1765)
London, Fritz (1900–1954)
London, Heinz (1907–1970)
Lorentz, Hendrik Antoon (1853–1928)
Lovelace, Ada Augusta [Countess of] (1815–1853)
Lucretius, Titus Carus (ca 99–ca 55 BC)
Lyapunov, Alexander Mikhailovich (1857–1918)
Mandelshtam, Leonid Isaakovich (1879–1944)
Marconi, Guglielmo (1874–1937)
Marcus Marci, Johannes (Jan) (1595–1667)
Mariotte, Edmé (1620–1684)
Martynov, Leonid Nikolaevich (1905–1980), Russian poet
Maxwell, James Clerk (1831–1879)
Mayer, Julius Robert von (1814–1878)
Meissner, Walter Fritz (1882–1974)
Mills, Robert Laurence (1927)
Misener, A.D.
Miura, Robert M. (1938)
Mollenauer, Linn Frederick (1937)
Monet, Claude (1840–1926), French painter
Morse, Samuel (1791–1872)
Mott, Nevill Francis (1905)
Mozart, Wolfgang Amadeus (1756–1791), German composer
Müller, Johannes Peter (1801–1858)
Müller, Wilhelm (1794–1827), German poet
Nakano, Tadeo (1926)
Navier, Louis Marie Henri (1785–1836)
Ne'eman, Yuval (1925)
Nekrasov, Aleksandr Ivanovich (1883–1957)
Neumann, John (Janos) von (1903–1957)
Newton, Isaac (1643–1727)
Niemi, Antti J.
Nishijima, Kazuhiko (1926)
Novikov, Sergei Petrovich (1938)
Nye, John Frederick
Ochsenfeld, R.
Oersted, Hans Christian (1777–1851)
Ohm, Georg Simon Christian (1787–1854)
Onnes, Heike Kammerlingh (1853–1926)
Onsager, Lars (1903–1976)
Ostrogradskii, Mikhail Vassilievich (1801–1862)

Parsons, William (Lord Ross) (1800–1867)
Pascal, Blaise (1623–1662)
Pasta, John R. (1918)
Pasternak, Boris Leonidovich (1890–1960), Russian poet
Pasteur, Louis (1822–1895)
Pauli, Wolfgang (1900–1958)
Petrovskii, Ivan Georgievich (1901–1973)
Petviashvili, Vladimir Iosifovich
Piskunov, N.S.
Pitaevskii, Lev Petrovich (1933)
Planck, Max Karl Ernst Ludwig (1858–1947)
Poe, Edgar Allan (1809–1849), American poet and essayist
Poincaré, Henri (1854–1912)
Poisson, Siméon Denis (1781–1840)
Polyakov, Aleksandr Markovich (1945)
Pontryagin, Lev Semenovich (1908–1988)
Popov, Aleksandr Stepanovich (1859–1906)
Popper, Karl Raimund [Sir] (1894–1963), English philosopher
Poynting, John Henri (1852–1914)
Pushkin, Aleksandr Sergeevich (1799–1837), Russian poet
Raman, Chandrasekhara Venkata (1888–1970)
Ramón-y-Cajal, Santiago (1852–1934)
Rankine, William John Macquorn (1820–1872)
Rayleigh, [Lord] John William (1842–1919) (J.W. Strutt)
Reynolds, Osborne (1842–1912)
Riemann, Georg Friedrich Bernhard (1826-1866)
Rubakov, Valerii Anatolievich (1955)
Russell, John Scott (1808–1882)
Rutherford, Ernest (1871–1937)
Sagdeev, Roald Zinnurievich (1932)
Saint-Venant, Adhemar Jean-Claude de (1797–1886)
Sakata, Shoichi (1911–1970)
Sakharov, Andrei Dmitrievich (1921–1989)
Salam, Abdus (1926–1996)
Savart, Felix (1791–1841)
Scherk, Joel
Schilling, Pavel L'vovich (1786–1837)
Schrieffer, John (1931)
Schrödinger, Erwin (1887–1961)
Schwarz, Albert Solomonovich
Schwarz, John H.
Schwinger, Julian Seymour (1918–1994)
Scott, Alwyn C.
Sechenov, Ivan Mikhailovich (1829–1905)
Seiberg, Nathan
Shabat, Aleksei Borisovich
Shelley, Percy Bysshe (1792–1822), English poet
Shubnikov, Lev Vassilievich (1901–1945)
Skyrme, Tony Hilton Roy (1922–1987)

Smith, K.
Sollogub, Fyodor Kuz'mich (1863–1927), Russian poet and writer
Soloviev, Vladimir Sergeevich (1853–1900), Russian philosopher and poet
Sommerfeld, Arnold Johannes Wilhelm (1868–1951)
Stalin, Josef Vissarionovich (1879–1953)
Steuerwald, Rudolf
Stokes, [Sir] George Gabriel (1819–1903)
Stolen, R.H.
Strutt, John William (see [Lord] Rayleigh)
Sturm, Jacques Charles François (1803–1855)
Tait, Peter Guthrie (1831–1901)
Tamm, Igor Evgenievich (1895–1971)
Tappert, Frederick D.
Thomson, [Sir] George Paget (1892–1975)
Thomson, [Sir] Joseph John (1856–1940)
Thomson, William (Lord Kelvin)
't Hooft, Gerardus (1946)
Tisza, Laszlo (1907)
Toda, Morikazu (1917)
Tomonaga, Sin-itiro (1906–1979)
Turing, Alan Mathison (1912–1954)
Tyutchev, Fyodor Ivanovich (1803–1873), Russian poet
Ulam, Stanislav Marcin (1909–1984)
Umov, Nikolai Alekseevich (1846–1915)
Verrier, Urbain Jean Josef Le (1811–1877)
Volta, Alessandro (1745–1827)
Voltaire (Arouet), François Marie (1694–1778)
Watt, James (1736–1819)
Wheatstone, Charles (1802–1875)
Weber, Eduard (1806–1871)
Weber, Ernst Heinrich (1795–1878)
Weber, Wilhelm Eduard (1804–1891)
Weierstrass, Karl Theodor Wilhelm (1815–1897)
Weinberg, Steven (1933)
Weiss, Pierre (1865–1940)
Whitehead, Alfred North (1861–1947), American philosopher
Wigner, Eugene Paul (1902–1995)
Witt, Aleksandr Adolfovich (?–1937)
Witten, Edward (1950)
Yang, Chen Ning (1927)
Young, Thomas (1773–1829)
Young, John Zachary (1907)
Yukawa, Hideki (1907–1981)
Zabusky, Norman J. (1928)
Zakharov, Vladimir Evgenievich (1939)
Zeldovich, Yakov Borisovich (1914–1987)
Zeno [of Elea] (ca 490–ca 430 BC)
Zumino, Bruno (1923)
Zweig, George (1937)

Index